The
Complete Book of
Locks & Locksmithing

Bill Phillips

Fourth Edition

McGraw-Hill, Inc.

New York San Francisco Washington, D.C. Auckland Bogotá
Caracas Lisbon London Madrid Mexico City Milan
Montreal New Delhi San Juan Singapore
Sydney Tokyo Toronto

Library of Congress Cataloging-in-Publication Data

Phillips, Bill, 1960-
    The complete book of locks & locksmithing /by Bill Phillips.–
4th ed.
      p.   cm.
    Includes index.
    ISBN 0-07-049866-0
    1. Locksmithing.   I. Title.
TS520.P54   1995
683'.3–dc20

95-9206
CIP

pbk 1 2 3 4 5 6 7 8 9 FGR/FGR 9 0 0 9 8 7 6 5

ISBN 0-07-049866-0

The editors of this book were: Acquisition editor, April Nolan; Executive editor, Joanne Slike; Managing editor, Lori Flaherty; and Manuscript editor, Michilinda Kinsey. The production supervisor was Katherine G. Brown. This book was set in ITC Century Light. It was composed in Blue Ridge Summit, Pa.

Printed and bound by Quebecor Printing, Fairfield, PA.

McGraw-Hill books are available at special quantity discounts to use as premiums and sales promotions, or for use in corporate training programs. For more information, please write to the Director of Special Sales, McGraw-Hill, 11 West 19th Street, New York, NY 10011. Or contact your local bookstore.

Product or brand names used in this book may be trade names or trademarks. Where we believe that there may be proprietary claims to such trade names or trademarks, the name has been used with an initial capital or it has been capitalized in the style used by the name claimant. Regardless of the capitalization used, all such names have been used in an editorial manner without any intent to convey endorsement of or other affiliation with the name claimant. Neither the author nor the publisher intends to express any judgment as to the validity or legal status of any such proprietary claims.

In order to receive additional information on these or any other McGraw-Hill titles, in the United States please call 1-800-822-8158. In other countries, contact your local McGraw-Hill representative.

MH95
0498660

Contents

Acknowledgments

This book could not have been written without the help I received from hundreds of individuals and companies. While there are far too many to mention each by name, I would like to single out the following companies that provided a great deal of information and materials:

A-1 Security Manufacturing Corp.; Abus Lock Co.; Adams Rite Manufacturing Co.; Alarm Lock Systems, Inc.; Arrow Lock; Belwith International Ltd.; Black & Decker U. S. Power Tools Group; CCTV Corp.; Dominion Lock Co.; Folger Adam Company; Framon Mfg. Co., Inc.; Ilco Unican Corporation; the International Association of Home Safety and Security Professionals, Inc.; Keedex Mfg.; Kustom Key, Inc.; Kwikset Corporation; Lock Corporation of America; M.A.G. Eng. & Mfg., Inc.; Master Lock Company; M.K. Morse Co.; Medeco Security Locks; Milwaukee Electric Tool Corp.; Monarch Tool and Mfg. Co.; The National School of Locksmithing and Alarms; Porter Cable Corp.; Preso-Matic Lock Co. Inc.; R&D Tool Company; Rofu International Corporation; Schlage Lock Company; Securitech Group, Inc.; Securitron; Security Engineering, Inc.; Sentry Door Lock Guards; Sentry Group; Simplex Access Controls Corp.; Skil Corporation; Slide Lock Tool Co.; Star Key & Lock Mfg.; Southern Steel Company; Taylor Lock Co.; Trine Consumer Products Division, Square "D" Co.; and Vaughan & Bushnell Manufacturing Co.

Also, special thanks to Bert Michaels, April Nolan, and Kimberly Tabor.

Introduction

Since 1976 when the first edition of *The Complete Book of Locks and Locksmithing* was published, it has become the world's best-selling locksmithing book. The various editions have been used as textbooks and reference sources by more students, apprentices, and professional locksmiths than any other book on the subject.

Although this 4th Edition is the most significant revision since 1976, it continues the tradition of being practical, comprehensive, and easy to understand. This 4th Edition has been redesigned to be more useful for locksmithing students, experienced locksmiths, and do-it-yourselfers who just want to make their home, office, or car more secure.

Each page has been updated, hundreds of photos and drawings have been added, and a comprehensive self-test (and answers) is included. The illustrations show locks and other devices from a variety of perspectives including outside views; inside views (what you would see if you had x-ray vision); cutaway views (outside views with a section cut away so you can see inside at an important area); and exploded views (showing all inside and outside parts side by side so you can see how they fit together.) The many photos and drawings make the book easy to use.

The written test in Chapter 22 is based on the International Association of Home Safety and Security Professionals' Registered Professional Locksmith (RPL) certification test. If you score well on the test, you should have no trouble passing an RPL certification test as well as other locksmithing certification and licensing tests.

No longer can a locksmith be competitive simply by cutting keys and installing and servicing locks. To be successful today, a locksmith must know about a wide range of security devices and systems including safes, burglar alarms, access control systems, home automation systems, and closed-circuit televisions. Today's locksmith must understand how the various devices and systems can be integrated to meet the needs of residential, commercial, and industrial customers. Even the locksmith who chooses not to install or service electronic security systems needs to be able to intelligently speak with such installers to most effectively integrate the locking devices and door hardware with other parts of a customer's security system.

The more a locksmith knows, the more jobs he or she will be able to perform or participate in. This 4th Edition is unique in two ways: it's written for the person who knows nothing about locksmithing, carpentry, or electricity, and it covers all the areas today's locksmith needs to know about, from choosing and using hand and power tools to starting and running a profitable locksmithing shop.

Many of the subjects covered in this 4th Edition aren't covered in other locksmithing books. Some subjects of special interest to professional locksmiths include selling, servicing, and installing safes, prison/detention devices, burglar alarms, car alarms, and door and window hardware. Other topics that may be of special interest to new locksmiths include masterkeying, lock picking, key impressioning, finding a job as a locksmith, starting a locksmithing business, and servicing automobile locks.

I could go on and on describing the tremendous amount of information contained in this book. After you've looked at the table of contents, flipped through the pages, and read a couple of chapters, no further explanation will be necessary.

I enjoyed writing this book and hope you enjoy reading it. Let me hear from you if you have any questions or comments. Send your comments to: Bill Phillips, c/o McGraw-Hill, Blue Ridge Summit, PA 17294-0850.

History of Locks and Locksmithing

Locks and keys have been around for centuries. They evolved with humankind as a natural part of our existence, signifying our continual need for protection of life and property. Every civilization in history understood the need for security, and such understanding spawned a persistent search for more sophisticated locks. In addition to providing security and peace of mind, locks and keys have often signified wealth and goodness. Religion and mythology attest to the role locks and keys played in early recorded history.

The Babylonian god Marduk was said to have made the gates and keys to Heaven. According to the Greeks, Hecate carried the keys to the universe, while the goddess Athena carried the keys of her city Athena. In Rome, a new bride was given keys at her wedding, but should she divorce, she must return them. In Matthew, Jesus said to Peter: "And I will give unto thee the keys of the kingdom of Heaven."

In Germany, the key was an amulet for protecting people against the Evil Eye. In China, an only son was given a key to lock him fast into life and living. During the Middle Ages through the eighteenth century, it was commonly held in Europe that anyone could block the consummation of a marriage by locking a lock or tying a knot with a cord and throwing the lock or cord into water during the marriage ceremony. Until the lock or cord was found and unlocked or unknotted, no real union of the married pair was possible.

Locks and keys were also believed to play an important role in childbirth and death. On the island of Salsette near Bombay, all locks on drawers and doors were unlocked to facilitate delivery. In Argyllshire, all locks in the house were unlocked during childbirth. In many parts of England it was believed that a person couldn't die as long as locks were locked or bolts were thrown in the house. When it was apparent that a sufferer was ready to die, it was a common practice to unlock all locks and slide all bolts to the open position to prevent unnecessarily prolonged agony.

Egypt

It isn't clear when the first lock was invented. The oldest known lock was unearthed in 1842 in Persia at Khorsabad. Paintings inside pyramids show that similar models were being made in Egypt about 2000 B.C. (Figure 1.1). The lock looked nothing like modern locks; it was large and made of hardwood. Like some of today's most sophisticated locks, however, it incorporated a bolt and pin tumblers, which gives testimony to the Egyptian genius.

The Ancient Egyptian's lock bolt was a solid horizontal beam hollowed at one end. The lock bolt was then mounted securely on the outside of a door or gate and passed through the lock body and into a bracket mounted on the adjoining wall. When in the locked position, hardwood or iron pins dropped into holes in the bolt, fixing it securely in the lock body and preventing the bolt from being moved to the unlocked position. The key was a length of wood with pegs, roughly the shape of a foot-long toothbrush. To unlock the lock, a person had to reach through an armhole (the forerunner of the modern keyhole) to insert the key through the hollow end of the bolt and raise the tumblers to the top surface of the bolt, freeing the bolt to slide to the unlocked position. As with locks today, the size, position, number, and length of the pin tumblers varied, which determines which keys can be used to open the lock.

Figure 1.1 An ancient Egyptian lock. To unlock the door, a large pegged key had to be positioned so that the key would lift the lock's pins to the shear line, which freed the horizontal bolt to be pulled to the open position.

Figure 1.2 Early Roman keys. Because Roman garments had no pockets, keys were made to be worn as finger rings.

Greece

Most early Greek doors pivoted at the center and were secured with rope tied in intricate knots (as in the Gordian knot of the Macedonian conqueror Alexander the Great). Some of the doors were secured by bolts from the inside. In the few cases where locks were used, they were very primitive. The key was a large crescent-shaped; it resembled a modern sickle. To work the bolt, the owner inserted the key into the keyhole, rotated it 180 degrees, and then pulled back to engage and raise wooden gravity pins to free the horizontal beam. A rope could then be used to slide the bolt back and forth, locking or unlocking the door.

Rome

The earliest known locks that look like models in use today are warded bit-key locks, which were invented by the Romans during the Metal Age. Originally the locks were iron and the keys were bronze. Because iron easily rusts and corrodes, few early Roman locks are in existence, but many early bit keys are available.

Wards are fixed obstructions at the keyway or within the lock case that block unauthorized keys from entering or operating a lock. In addition to using a wide variety of wards, the Romans made ingenious attempts to disguise the keyway as ornamentation or to hide it altogether. One famous Roman lock is in the shape of a fish. The keyway is revealed when one of the fins is turned.

The biggest shortcoming of wards is that no matter how many are contained in a lock and no matter how they are positioned, they can't prevent ultra-thin or "skeleton" keys from operating the lock. A bent piece of wire, for instance, can easily bypass all the wards of a typical bit-key lock.

Other Roman innovations included the metal key, the spring-loaded bolt, the spring-operated padlock, and the first true (i.e., removable) padlock. One curious aspect of Roman lock lore is that the keys were designed to be worn as rings because togas did not have pockets (Figure 1.2).

After the fall of Rome, Europe was swept by Barbarian hordes moving south and east from the Rhineland and Scandinavia. Brute force was the order of the day, and locks meant little. People put their faith in arms and fortresses.

This period passed as the Barbarians learned from their victims and settled into the routines of farming and fishing. The Vikings who settled in Brittany evolved a new kind of civilization, with a king at the apex and power radiating downward into various levels of nobility. This social pattern, known as feudalism, gave the nobility access to wealth and property, and the art of locksmithing again flourished.

Europe

In Europe, keys were made that could move about a post and shift the position of a movable bar (the locking bolt). In its various forms it worked quite well. The first obstacles to unauthorized use of the lock were various internal wards. Medieval and Renaissance craftsmen made warded locks and improved upon them. Some were very complex, using many interlocking wards and complicated keys. Since many of these wards could be easily bypassed by almost anyone, newer methods had to be devised.

Later, chests were made entirely of metal; the locksmith reasoned that if one bolt was secure, then a chest with eight, ten, twelve, or even fifteen locking bolts would be more secure. Thus, craftsmen created elaborate internal mechanisms to allow the many bolts to be shifted by one or two separate keys. Levers, springs, ratchets, and pinions were employed to do this job. The locksmiths also installed separate locks for separate bolts; the locks had false keyholes and safety devices, such as spring-loaded knives that would injure or kill the thief. Hidden keyholes were extremely popular among the merchant and upper classes.

In France, the treatise The Art of the Locksmith (1767) was published, describing examples of the tumbler or lever lock. Exactly who invented it is unknown, but credit is long overdue. As locksmithing advanced, locks came with two, three, six, and more levers. Each lever had to be lifted, and when all were in proper alignment, the bolt could be moved, opening or closing the lock.

England

The English also worked on newer and better locking devices. Incentive was given in the form of cash awards and honors to those who could successfully open these newer and more complex locks. In the forefront of lock designing were three Englishmen: Robert Barron, Joseph Bramah, and Jeremiah Chubb.

Robert Barron patented the double-acting tumbler lock in 1778, which was more secure than other locks during that time. Like other lever tumbler locks, Barron's lock employed wards. But it also used a series of lever tumblers, each of which was acted upon by a separate step of the key. If any tumbler wasn't raised to the right height by the key, its contact with a bolt stump would obstruct bolt movement (Figure 1.3). Barron's lock corrected the shortcoming of earlier lever tumbler locks, which could easily be circumvented by any key or instrument thin enough to bypass the wards. Barron added up to six of these double-lever actions to his lock and thought it virtually impossible to open except with the proper key. He soon found out differently.

Another Englishman, Joseph Bramah, wrote A Dissertation on the Construction of Locks, which exposed the many weaknesses of existing so-called thiefproof locks and pointed out that any of them could be picked by a good lock specialist or criminal with some training in locks and keys. Bramah admitted that Barron's lock had many good points, but he also revealed its major fault: the levers, when in the locked position, gave away the lock's secret. The levers had uneven edges at the bottom; thus, a key coated with wax could be inserted into the lock and a new key could be made by filing where the wax had been pressed down or scraped away. Several tries could create a key that matched the lock. Bramah pointed out that the bottom edges of the levers showed exactly the depths the new key should be cut in order to clear the bolt. Bramah suggested that the lever bottoms should have a smooth surface, and that the lever slots should be cut unevenly. Then only a master locksmith could open the lock.

Using those guidelines, Bramah patented a lock in 1784 that employed a series of notched lever tumblers that were aligned by corresponding notches in the key. The novelty was in the way the levers were arranged. Earlier locks were built on the pattern shown in Figure 1.3; Bramah's lever tumblers were mounted vertically. Bramah used radial tumblers and a barrel key (Figure 1.4). The pin on the side of the key rotated the lock. For about 50 years after its introduction, the Bramah lock was harder to pick than any other lock. The lock wasn't "pickproof," but picking it was a job beyond the average thief. Each tumbler had to be aligned, and the control piece had to be turned in the right direction.

During this period, burglary was a major problem. After the Portsmouth, England dockyard was burglarized in 1817, the British Crown offered a reward to anyone who could devise a lock that couldn't be picked open. A year later, Jeremiah Chubb patented his lock and won the prize money.

Chubb's lock had many improvements over Barron's. One of the improvements was a metal "curtain," which fell across the keyhole when the mechanism began to turn, making the lock difficult to pick. Chubb's lock also added a detector lever that indicated whether the lock had been tampered with. A pick or an improperly cut key would raise one of the levers too high for the bolt gate. This movement engaged a pin that locked the detector lever. The detector could be cleared by turning the correct key backwards and then forwards.

Chubb's lock got much attention. One convict, a former locksmith, claimed he could open the best of locks. He was guaranteed $250 and his freedom if he could open Chubb's lock. The British government and Chubb supplied the convict with a

Figure 1.3 An eighteenth century lever tumbler lock. The width of the tumblers corresponded to top slots in the key.

Figure 1.4 A Bramah radial lever lock (circa 1790).

lock, key blanks, appropriate lock picking implements, and some dismantled locks to work on. After working for some three months, he finally gave up and served out his sentence. England announced that it finally had an unpickable lock.

But in 1851 at the International Industrial Exhibition in London, an American locksmith named Alfred C. Hobbs picked open both the Brahmah and the Chubb locks in less than half an hour. Hobbs patented what he called "Protector" locks. In 1854, one of Chubb's locksmiths used special tools to pick open one of Hobb's locks.

American Locks

Early American lock bolts were mounted on the inside of the door and could be opened from the outside by means of a latchstring. Hence the phrase, "the latchstring's always out." At night, the string would be pulled inside, "locking" the door. Of course, someone had to be inside to release the bolt. An empty house was left unlocked. As the country was settled, theft increased, and local merchants and blacksmiths soon got into the lock trade. Small cabinet and cupboard locks were very popular. The lady of the house (or, in large establishments, the butler) was in charge of the household keys. The doors were unlocked in the morning and the cabinets and cupboards were opened to dispense the day's supplies and immediately relocked. At night the doors were secured.

As industry progressed in the United States, there was an increasing demand for new and better locks. Locksmiths responded by copying and improving on the English locks. In 1836, a New Jersey locksmith named Solomon Andrews developed a lock with adjustable tumblers and keys, which allowed the owner to rekey the lock any time. Because the key itself could also be modified, there was no need to use a new key to operate a rekeyed lock. But few homeowners used the lock because rekeying it required dexterity, practice and skill. The lock was of more interest to banks and businesses.

In the 1850s, two inventors, Andrews and Newell, were granted patents on an important new feature-removable tumblers. The tumblers could be disassembled and scrambled to make, in effect, a new lock. The keys had interchangeable bits that matched the various tumbler arrangements. After locking up for the night, a prudent owner would scramble the key bits. Even if a thief got possession of the key, it would take him hours to stumble on the right combination. In addition to removable tumblers, this lock featured a double set of internal levers.

Newell was so proud of this lock that he offered a reward of $500 to anyone who could open it. A master mechanic took him up on the offer and collected the money. This experience convinced Newell that the only secure lock would have its internals sealed off from view. Ultimately, sealed locks appeared on bank safes in the form of combination locks.

As we have seen, Hobbs picked the famed English locks with ease. Until Hobbs' time, locks were opened by making a series of false keys. If the series was complete, one of the false keys would match the original. Of course, this procedure took time. Thousands of hours might pass before the right combination was found. Hobbs depended upon manual dexterity. He applied pressure on the bolt while manipulating one lever at a time with a small pick inserted through the keyhole. As each lever tumbler unlatched, the bolt moved a hundredth of an inch or so.

Until the early nineteenth century, locks were made by hand. Each locksmith had his own ideas about the type of mechanism—the number of lever tumblers, wards, and internal cams to put into a given lock. Keys contained the same individuality. A lock could have 20 levers and weigh as much as five pounds.

In 1805, an American physician named Abraham O. Stansbury was granted an English patent for a pin tumbler lock that had been based on the principles of both the ancient Egyptian and Brahmah locks. Two years later, his design was granted the first lock patent by the U. S. Patent and Trademark office. Stansbury's lock used segmented pins that automatically relocked when any tumbler is pushed too far. The double-acting pin tumbler lock was never manufactured for sale.

In 1844, Linus Yale, Sr., of Middletown, Connecticut, patented his "Quadruplex" bank lock, which incorporated a combination of ancient Egyptian design features and mechanical principles of the Brahmah and Stansbury locks. The Quadruplex had a cylinder subassembly that denied direct access to the lock bolt.

In 1848, Yale patented another pin tumbler design based on the design of the ancient Egyptian lock and the Brahmah lock. But his son, Linus Yale, Jr., improved on the lock design and is credited with inventing the modern pin tumbler lock. Figure 1.5 illustrates the working principle of the lock cylinder.

Patented in 1865, Yale, Jr.'s pin tumbler mortise cylinder may be the most important development in locks since the ancient Egyptian lock was invented. Not only could Yale, Jr.'s lock easily be rekeyed, but it provided a high level of security, could

Figure 1.5 The operation of the Yale pin tumbler cylinder. The key raises the tumblers to the shear line, which frees the plug to be rotated to the locked or unlocked position.

easily be mass produced, and could be used on doors of various thicknesses. His lock design meant that keys no longer had to pass through the thickness of a door to reach the tumblers or bolt mechanism, which allowed keys to be made smaller and thinner. (Linus Yale Jr.'s first pin tumbler locks were designed for drawers, banks, and doors. They were marked L. Yale Jr., along with the 1865 patent date. A big difference between his pin tumbler locks and modern versions is that he used a flat steel key rather than a paracentric cylinder key.)

Since 1865, few major changes have been made to the basic design of mechanical lock cylinders. Most cylinder refinements since that time have been limited to using unique keyways (along with corresponding shaped keys), adding tumblers, varying tumbler positions, varying tumbler sizes and shapes, and combining two or more basic types of internal construction, such as the use of both pin tumblers and wards.

Other important lock innovations involved external design and operating technology. In 1833, three brothers, called the Blake Brothers, were granted a patent for a unique door latch that had two connecting doorknobs. It was installed by boring two connecting holes. The larger hole, which was drilled through the door face, was for the knob mechanism. The smaller hole, which was drilled from the door edge, was for the latch. In 1834, the brothers, formed the Blake Brothers Lock Company to produce and sell their unusual latch. The big difference between their latch and others of their time was that all the door locks were installed by being surface mounted to the inside surface of a door. At that time, the brothers probably never imagined that 100 years later their development would be used to revolutionize lock designs.

In 1923, Walter Reinhold Schlage of San Francisco was granted a patent (No. 1,456,041) for the first cylindrical shaped door lock, which was based on the Blake Brothers' door latch. The patent described a lock that consisted of a knob, spindle, and latch retractor all in one self-contained unit within a circular housing. The lock was the first of many versions of Schlage Lock Company's "A" locks. The first versions of the lock didn't use a key. The first key-operated "A" locks used wafer (or disc) tumblers. In 1982, the wafer tumbler versions were discontinued. Today, most cylindrical locks (more commonly called key-in-knob locks) use pin tumbler cylinders, and they are possibly the most popular type of door lock used on American homes.

Another significant American lock development was the time lock, devised in 1872 by James Sargent of New York State. Sargent's design was based on an earlier Scottish patent. It isn't clear when the first electric lock was developed, but between 1887 and 1900 various American inventors received patents for dozens of electric locking devices.

(For historical information about current lockmakers, see Appendix A.)

Early American lock companies

In 1832, the F. T. Stanley Company of New Britain, Connecticut became the first lock company in America. In 1833, locks were being made in Terryville, Connecticut in a building housing the Eli Terry Clock Shop. The lock making business was started by Stephen G. Bucknall, a locksmith from England. He worked for a firm called Lewis, McKee and Company. The company changed hands, first becoming Lewis & Gaylord, and then Lewis Lock Company. Bucknall went on to help form a company called Bucknall, McKee and Company, which eventually became Williams, McKee and Company. In 1846, the company was sold to James Terry and his associates.

The James Terry Company and the Lewis Company merged to create Eagle Lock Company in 1854, and Terry served as president. In 1889, the Eagle Lock Company built a new office building two blocks from the Eli Terry Clock Shop. (The company's original offices were sold to the town and used as the Plymouth Town Hall from 1890 to 1975, when a new town hall was built. Then the old Eagle Lock Company building was torn down. Today, the Lock Museum of America is located at the site.) Eagle Lock Company stopped doing business in 1976, ending 122 years of lockmaking.

Linus Yale, Sr. owned the Yale Lock Shop in Newport, New York until his death in 1857. Then Linus Yale, Jr. moved to Philadelphia and formed the L. Yale Jr. Company to make bank locks. In 1861, he moved his shop to Massachusetts and teamed with Colonel Halbert Greenleaf to form the Yale and Greenleaf Manufacturing Company. In 1868, Yale teamed with Henry R. Towne to form the Yale Lock Manufacturing Company. Shortly after its formation, the company was moved to Stamford, Connecticut. Linus Yale., Jr, died of a heart attack on December 25, 1868. His partner, Henry Towne, took over and built the company into a manufacturing giant.

The oldest active lock manufacturer in America, the Russell & Erwin Company (now Corbin Russwin) began in Connecticut in 1839.

The Role of Locksmiths

As locks have changed over the years, so too has the role of locksmiths. Early locksmiths were part craftsmen and part artist; they made exquisite locks and keys by hand. They developed their skills by going through an apprentice system, which eventually became known as The Guild System. The system required a would-be locksmith to work as an apprentice for a master locksmith. The apprentice then had to pass a rigid examination to be considered a journeyman locksmith. To become known as a master locksmith, the journeyman had to create a lock (called a master-piece) that a master locksmith judged to be of high quality.

Today, locks and keys are mass produced and few locksmiths help create them. Modern locksmiths are part mechanic, part carpenter, part electrician, and part security consultant. Basic tasks they perform include: selling, installing, and servicing mechanical and electrical locking devices; cutting keys; gaining entry to locked buildings and automobiles; and providing security advice. Those things are done by virtually all locksmiths. Many locksmiths also have specialized knowledge of a wide range of security devices and systems—such as safes, burglar alarms, car alarms, and closed circuit television systems.

Although The Guild System is no longer active, some people still enter the trade through serving as an apprentice. Many people also learn basic locksmithing skills by going to school, taking a home-study course, or by studying a book such as this one. How a person enters the trade is less important than how skillful the person is. Because the trade is constantly evolving, locksmiths must continually improve and expand their skills. That's why it's important for locksmiths to subscribe to trade journals, join trade associations, and attend training classes. (To learn more about current requirements for becoming a locksmith see Chapter 20.)

2

Tools of the Trade

In order to make money in locksmithing, you have to work quickly and professionally, which requires using proper tools and supplies. In addition to many of the hand and power tools commonly used by carpenters and electricians, locksmiths also use a variety of special tools. The special items needed depends on which locksmithing tasks you will perform.

The more services you want to perform, the more special tools you'll need. Quality is often more important than quantity. A tool that breaks, bends, or stops working at the wrong time can cause you to lose a lot of money. On the other hand, a tool that's designed to last a lifetime will pay for itself many times. Even if you have little money to invest in tools, always get the highest quality products that you can afford.

This chapter lists all the basic tools and supplies you'll need to perform any locksmithing task. It also tells what to look for when buying tools and supplies.

Electric Drills

It's important to choose the right drill because it's used by locksmiths more often than any other type of portable power tool (Figure 2.1). Drills range in price from under $50 to more than $200. Although price is often an indication of the quality of a drill, the only way to ensure that you're getting the most for your money is to know the important ways drills differ from one another.

Three common drill sizes are available: ¼-inch, ⅜-inch, and ½-inch. Drill size is determined by the largest diameter drill bit shank the drill's chuck can hold without an adapter. For example, a drill whose chuck can hold a drill bit shank of up to a ½ inch is called a *½-inch drill*.

A drill's power is a combination of chuck speed and torque. Speed is measured in rpm (revolutions per minute when the chuck is spinning freely in the air). *Torque* refers to the twisting force at the chuck when the drill is being used to drill a hole.

Chuck

Chuck jaws
(In closed position)

Forward/reverse
switch

Trigger

Figure 2.1 An electric drill is one of the most useful tools a locksmith can have. (*Black & Decker*)

Any drill will slow down while drilling a hole, but the more torque it has the easier it will bore through material.

Chuck speed and torque are largely determined by the type of *reduction* gears a drill has. Reduction gears in a drill work somewhat like car gears. One gear, for example, lets the car move quickly on a flat road. Another gear provides the car more power when climbing a steep hill. This analogy isn't perfect, however, because a drill comes with a fixed type of reduction gear set. You can't shift the gears of a drill.

A *single-stage* reduction gear set lets a drill's chuck spin extremely fast in the air but supplies little torque. A *two-stage* reduction gear set provides fewer rpms, and more torque. A *three-stage* reduction gear set further reduces rpms and increases torque. Greater torque is especially useful when drilling hardwood, steel, or other hard materials. It's also useful when drilling large holes, such as those needed to install deadbolt locks.

Most ¼-inch drills have a single-stage reduction gear set. Their chucks commonly spin at 2500 rpms or more. Such drills are lightweight and most often used for drilling plastic, softwood, and sheet metal. Using a ¼-inch drill to drill hardwood or steel would be time-consuming and could damage the drill.

Most ⅜-inch drills have either a single-stage or two-stage set of reduction gears. The shanks of many popular hole saws and drilling accessories won't fit into a ⅜-inch chuck without using an adapter.

The ½-inch drill is by far the most popular size among locksmiths. A ½-inch drill usually has a two- or three-stage set of reduction gears, and its chuck typically spins at up to 600 rpms. The ½-inch can drill any material that can be drilled by a smaller drill.

Not all ½-inch drills are alike. They can differ greatly in quality and price. Many manufacturers call their drills "heavy-duty," "professional," "commercial," and the like. Such labels are primarily for promotional purposes; they have no industry standard meanings. It's best to look beyond such terms for specific features. The most important features are two- or three-stage reduction gears, variable speed reversing, double insulation, antifriction bearings, and at least four amps.

A *variable speed* feature provides maximum control over drilling speed, allowing you to drill different materials at different speeds. Without variable speed control, you might be limited to drilling at one or two speeds only. Many drills also have a switch that lets you reverse the direction that the chuck turns. That is useful for backing out screws or a stuck drill bit. A drill that has a variable speed feature and is reversible is called *variable speed reversible* (VSR). VSR drills are well worth the few extra dollars they cost.

Double insulation means the drill is housed in nonconductive material (such as plastic) and the motor is isolated from other parts of the drill by a nonconductive material. Double insulation protects you from getting shocked. Most high-quality drills are double insulated. Don't mistake plastic housing as a sign of low-quality. All-metal housing is a sign of low-quality.

Antifriction (*ball* or *needle*) *bearings* help a drill run smoothly and last long. All high-quality drills use some antifriction bearings. Low-quality models use all plain sleeve bearings.

An *amp* (short for *amperes*) is a unit of electric current. In general, the more amps used the more powerful the drill. High-quality drills usually use at least four amps. (Drills that list horsepower rather than amps are usually low-quality.)

Any drill that has all of these features is a high-quality model that should last a lifetime if it's properly used and maintained. Some locksmiths have additional considerations such as a drill's color, size, weight, feel, brand, and country of origin. Those additional considerations are personal and have little to do with a drill's quality.

Cordless Drills

A cordless drill can conveniently insert screws and drill small holes (Figure 2.2). It's typically lighter than an electric drill and can be used in places with no electrical outlet. As a rule, cordless drills are less powerful than their electric counterparts. Another problem with cordless drills is that their batteries have to be recharged and replaced. Although a cordless drill can be great as a backup or extra drill, it isn't a good choice if you can afford only one drill.

Key Cutting Machines

Because cutting keys is often a major source of income for a locksmith, *key cutting* (or *duplicating*) *machines* are among the most important tools a locksmith buys.

Figure 2.2 A cordless drill can be useful when no electricity is available. (*Milwaukee Tool Co.*)

Expensive machines often can cut a wide variety of keys for standard and high-security locks. Two popular types of key cutting machines are shown in Figure 2.3.

If you have little money to invest in key machines, it's probably best to start with a low-cost model that cuts cylinder keys—the most common type of keys for home and car locks. Later, you might want to add a low-cost machine that cuts flat keys (such as those used for safe deposit boxes) or tubular keys (such as those used for

Figure 2.3 Different types of key machines are needed to cut cylinder keys (left picture) and tubular key lock keys (right picture).

vending machines and laundromat equipment). Another way to save money is to buy only portable dual-voltage models that can be used either in the shop or in a service vehicle.

Workbench

Whether you practice locksmithing in a shop or at home, you'll need a workbench. You can use a table or desk on a temporary basis, but a workbench is more practical and comfortable. If you have basic woodworking tools and skills, you should have no trouble making your own workbench.

The workbench should be:

- Long enough to ensure adequate work space
- Strong enough to support a key machine at one end, out of the way
- Solid enough to keep the key machine in alignment
- High enough to allow you to work without stooping
- Wide enough to store parts and supplies; 30 inches is the comfortable maximum
- Lit from overhead and behind or from the sides (never have the light too close to the bench)

Workbench location

Consider where you want to locate your workbench. If possible, place it in a well ventilated area, away from general traffic, near frequently used equipment and supplies, with an easy access from several directions. Many locksmiths place storage bins near their workbenches to hold tools and supplies (Figure 2.4).

Figure 2.4 Bins can be helpful for sorting lock parts and supplies.

Figure 2.5 A simple bench is adequate for a student locksmith.

Figure 2.6 Benches with a full complement of drawers and partitioned overhead bins are ideal for professional locksmiths.

Examples of workbench designs are shown in Figures 2.5 and 2.6. No one type of workbench is ideal for everybody; individual tastes and needs differ.

Install a vise at one end, out of the way. Mount the vise on a swivel base, sturdily built, with jaws at least 3 inches wide. The jaws should open to at least 5 inches.

List 1: Common hand and power tools all locksmiths should have:

- Allen wrench set
- Bench grinder with wire wheel
- Bolt cutters, 16-inch
- C-clamps (Figure 2.7)
- Center punch set
- Code books, general set
- Combination square
- Coping saw and blades

- Dent puller
- Disc grinder
- Drill bits, auger assortment
- Drill bits, expansion assortment
- Drill bits, masonry assortment (Figure 2.8)
- Drill bits, spade assortment
- Drill bits, straight assortment
- Drill, cordless
- Drill, electric with ½-inch chuck
- Extension cord, 50-foot
- Files, assorted types and sizes (Figure 2.9)
- Flashlight
- Hacksaw and blades (Figure 2.10)
- Hammers: claw, ball-peen, and soft-face (Figure 2.11)
- Hand cleaner
- Hollow mill rivet set
- Lever, carpenter's (from 18 to 96 inches long with three vials)
- Lever, torpedo (up to 9 inches long)
- Lubricant (such as WD-40)
- Mallet, rubber
- Masking tape
- Nails and screws, assorted
- One-way screw removal tool
- Paint scrapper
- Pencils
- Pliers, adjustable
- Pliers, cutting
- Pliers, long nose (7-inch)
- Pliers, locking
- Pliers, slip-joint
- Retractable tape measure, 25-foot
- Rivet assortment
- Safety glasses
- Scratch awl
- Sandpaper and emery cloth
- Screwdriver bits, assorted Phillips and slotted
- Screwdrivers, assorted Phillips and slotted (Figure 2.12)
- Scissors for paper
- Snap ring pliers, assorted sizes

- Socket sets, ½-inch and ¼-inch
- Storage trays (Figure 2.13)
- Tap set
- Toolboxes
- Wrenches, adjustable and pipe

Figure 2.7 C-clamps usually range in size from 1 to 8 inches.

Figure 2.8 Masonry bits used for drilling holes in brick and concrete. (*Keedex Mfg.*)

Figure 2.9 Files can be useful for cutting keys by hand and performing many other locksmithing tasks.

Figure 2.10 A hacksaw for cutting metals and other nonwood materials.

Straight claw or rip hammer
For use with non-hardened common or finishing nails only. Choose 20,24,28, or 32 oz. weights for framing and ripping.

Ball pein hammer
For riveting, center punching and bending or shaping soft metal. Choose weights from 2 oz. to 48 oz., depending on the task.

Hand drilling hammer
Easy to handle with plenty of punch. The only hammer to use with star drills, masonry nails, steel chisels and nail pullers

Brick hammer
Designed for cutting and setting bricks or blocks, and for chipping mortar.

Curved claw or nail hammer
For use with non-hardened common or finishing nails only. Choose 16 or 20 oz. weights for general carpentry, lighter weights for model work or fine cabinetry.

Rawhide mallet
For use in furniture assembly, shaping soft sheet metals or any project that requires non-marring blows. Choose weights from 3oz. to 24 oz.

Shingling hatchet
The Pro's choice for installing shakes and shingles. Handy, three position gauge pin for metric and U.S. standards sets exposed length of shingle. Milled and crowned face sets nails cleanly.

Wallboard tool
Multi-purpose tool scores wallboard, mistakes cutouts and sets nails with a perfect dimple. Exclusive "Flat Top" head permits nailing in corners.

Figure 2.11 Striking tools are useful to locksmiths. (*Vaughan & Bushnell Manufacturing Company*)

Figure 2.12 Electrician's screwdrivers (top) are designed for precision work; standard screwdrivers are used on heavier jobs (bottom). (*Stanley Works*)

Figure 2.13 Parts and storage boxes are needed for work in and out of the shop.

List 2: Tools for shop/bench work (needed in addition to all of the tools in List 1)

- Code books
- Cylinder cap remover
- Dial caliper (Figure 2.14)
- Flat steel spring stock
- Interchangeable core capping machine (Figure 2.15)
- Interchangeable core lock service kit (Figure 2.16)
- Key blanks, assorted
- Key cutting machine, electric
- Key duplicating machine, electric
- Key marking tools
- Lock parts, assorted
- Lock pick gun
- Lock pick set (Figure 2.17)
- Lock reading tool
- Mortise cylinder clamp (Figure 2.18)
- Pin kit (Figure 2.19)
- Pin tray
- Pin tumbler tweezers (Figure 2.20)
- Plug followers, assorted sizes (Figure 2.21)
- Plug holders (Figure 2.22)
- Plug spinner (Figure 2.23)
- Retainer ring assortment
- Round spring steel, assorted sizes
- Shim stock (Figure 2.24)
- Spindle assortment
- Spring assortment
- Tension wrenches (Figure 2.25)
- Tubular-key decoder (Figure 2.26)
- Tubular-key lock picks (Figure 2.27)
- Tubular-key lock saw (Figure 2.28)
- Vise
- Whisk broom
- Workbench

Figure 2.14 A dial caliper allows locksmiths to accurately measure pin tumblers, key blanks, key cut depths, and plugs. (*Ilco Unican Corp.*)

Figure 2.15 An interchangeable core capping machine is needed for capping IC cores. (*Arrow Mfg. Company*)

Figure 2.16 An interchangeable core cylinder and key stamping fixture and supplies needed for servicing IC cores. (*Arrow Mfg. Company*)

Diamond

Double Ball

Ball

Riffle

Hook

Figure 2.17 A variety of lock picks allow the locksmith to choose the best one for a given lock. (*A-1 Security Manufacturing Corp.*)

Figure 2.18 A cylinder removal tool makes it easy to remove stubborn mortise cylinders without stripping their threads. (*A-1 Security Manufacturing Corp.*)

Figure 2.19 A pin kit holds tumblers, springs, and other supplies needed to rekey lock cylinders. (*Arrow Mfg. Company*)

Figure 2.20 Pin tumbler tweezers make it easy to handle tumblers and other small parts. (*Ilco Unican Corp.*)

Figure 2.21 Plug followers come in various sizes for different size plugs. The devices prevent top pins and springs from falling into the cylinder while the bottom pins are being replaced. (*Ilco Unican Corp.*)

Figure 2.22 Plug holders come in various sizes; some models can hold several plugs of different sizes. In addition to holding a plug that's being serviced, plug holders allow a locksmith to make sure a rekeyed plug will rotate in the cylinder properly before inserting it back into the cylinder. (*A-1 Security Manufacturing Corp.*)

Figure 2.23 A plug spinner is designed to be inserted into a keyway and to use spring pressure to quickly rotate a plug clockwise or counterclockwise. The device is used when a lock has been picked in the wrong direction. (*A-1 Security Manufacturing Corp.*)

Figure 2.24 Shims are used when disassembling pin tumbler cylinders. (*Ilco*)

Figure 2.25 Tension wrenches come in various shapes and sizes. They're designed to partially enter a keyway along with a lock pick, so that when all the tumblers are at the shear line the tension wrench can rotate the cylinder as the proper key would. (*A-1 Security Manufacturing Corp.*)

Figure 2.26 A tubular key decoder makes it easy for a locksmith to determine the bitting of tubular keys. (*A-1 Security Manufacturing Corp.*)

Figure 2.27 A tubular key lock pick is needed to pick tubular key locks. (*A-1 Security Manufacturing Corp.*)

Figure 2.28 A tubular key lock saw is used to drill through the tumblers of tubular key locks. (*A-1 Security Manufacturing Corp.*)

List 3: Tools for automotive work (needed in addition to all of the tools in Lists 1 and 2)

- Automobile entry tools and wedges
- Bezel nut wrench
- Broken-key extractors (Figure 2.29)
- Chrysler shaft puller
- Code books, automotive
- Code key cutting machine, manual (Figure 2.30)
- Door handle clip removal tool (Figure 2.31)
- Door trim pad clip removal tool (Figure 2.32)
- Face caps
- Face cap pliers
- Flexible light
- General Motors lock decoder (Figure 2.33)
- Lock plate compressor
- Steering column lock plate compressor
- Steering wheel pullers
- VATS/PASSKey decoder or key analyzer (Figure 2.34)

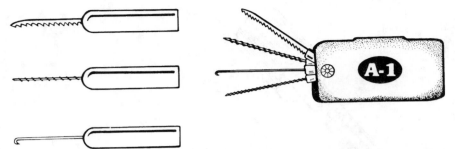

Figure 2.29 Broken key extractors are useful for removing broken key parts from locks. (*A-1 Security Manufacturing Corp.*)

Figure 2.30 A mechanical code key cutter lets locksmiths make accurate keys without having a key to duplicate. (*Ilco*)

Figure 2.31 A door handle clip tool helps remove the retainer clip that secures an automobile's door handle to the door. (*A-1 Security Manufacturing Corp.*)

Figure 2.32 A door trim pad clip removal tool is used for automotive work. (*A-1 Security Manufacturing Corp.*)

Figure 2.33 A General Motors Decoder decodes the tumblers in a GM lock without complete disassembly of the lock. (*Keedex Mfg.*)

Figure 2.34 A VATS/PASSKey decoder lets locksmiths quickly determine the electrical resistance values of VATS/PASSKey key lanks. (*Keedex Mfg.*)

List 4: Tools for servicing safes, vaults, and safe deposit boxes (needed in addition to all of the tools in List 1 and List 2)

- Borescope
- Carbide drill bits
- Change keys, assorted (Figure 2.35)
- Door puller
- Drill rig (Figure 2.36)
- Hammer drill
- Nose puller
- Safe-moving equipment
- Sledgehammer

Figure 2.35 Safe change keys are needed to change the combinations of certain types of safes. (*Keedex Mfg.*)

Figure 2.36 A drill rig holds a drill in place while a safe is being drilled. (*Keedex Mfg.*)

List 5: Tools/supplies for installing door locks and other door hardware (needed in addition to all of the tools in Lists 1 and 2)

- Boring jigs (Figure 2.37)
- Broom and dustpan
- Compass (or keyhole) saw
- Door reinforcers, assorted sizes and finishes
- Drop cloth
- Drywall (or wallboard) saw
- Filler plates
- Hole saws, assorted sizes (Figure 2.38)
- Kwikset cylinder removal tool
- Lever
- Nails, assorted types and sizes
- Pry bar
- Reciprocating saw
- Screw gun
- Screws, assorted sizes
- Screws, one-way
- Shovel
- Strike plates and strike boxes, assorted types and sizes
- Utility knife and assorted blades
- Vacuum cleaner
- Weiser shim pick
- Wood chisels, assorted sizes (Figure 2.39)
- Wood glue

Figure 2.37 A boring jig is a template that acts as a guide for drilling precise installation holes and cutouts for locks and door hardware. (*Keedex Mfg.*)

Figure 2.38 Hole saws and spade bits are used to install locks on doors. (*Skil Corporation*)

Figure 2.39 Wood chisels of various sizes are needed when installing locks and other hardware on wood doors. (*Stanley Works*)

List 6: Tools for installing alarms and electronic security devices (needed in addition to all of the tools in Lists 1 and 2)

- Electrical tape
- Fish tape
- Flexible drill bits and extensions
- Multimeter
- Staple guns (or wiring tackers) and staples, assorted sizes for wire and coaxial cable
- Twist connectors
- Under-carpet tape
- Voltage tester
- Wire stripper
- Wire cutters

It isn't necessary to have all of the listed tools to practice the lessons in this book. As you study each chapter, you'll learn which tools you need and which you can do without. With all the tools listed in this chapter, you'll have enough equipment and supplies to start a locksmithing business.

3

Types of Locks and Keys

Most people have a general understanding of what a lock is, but that understanding is often limited to the few models they've owned or seen. Few nonlocksmiths know much about the wide variety of locks that exist throughout the world or how locking technology has changed over the years. That's why many general consumer books and articles about locks (which are almost always written by nonlocksmiths) use definitions of the term *lock* that are too narrow, too broad, or too outdated. Although laypersons might not need to be concerned about the misleading definitions, it's important for a locksmith to know what a lock really is and isn't.

Merriam Webster's Collegiate Dictionary, Tenth Edition defines *lock* as: "a fastening (as for a door) operated by a key or a combination." Although the definition isn't wrong, it's too narrow. Some locks, such as automobile ignition locks, are used to turn things on and off (by completing and breaking an electric circuit) rather than to fasten. The dictionary's definition is also outdated, because many modern locks are operated by things other than keys or combinations, such as magnetically coded cards or voice patterns. The following overly broad description of *lock* was given by freelance writer Stephen Tchudi in his book *Lock & Key*:

> We can imagine that the first lockmaker was a person who returned from foraging or hunting to find his cave had been raided, perhaps by animals, possibly by people. To protect his property, he got the idea of placing a stone, the biggest one he could move, over the mouth of the cave, thus hiding the cave entrance and blocking—or locking—it against unwanted entry. He may have discovered that this "lock" worked well with animals, which couldn't figure out the "key" to get into the cave.
>
> In Ancient India the emperor of Annam is reported to have made a system of "living locks." He placed his valuable gold and jewels on small islands in a moat. Then he released crocodiles in the water. And he didn't feed the crocs all that often. So his "locks" were always on the prowl for a good meal. . . .

A lock, after all, is simply a barrier or closure, a way of sealing up an entryway, of keeping what you want in, *in*, what you want out, *out*. . . . Strictly speaking, hiding your money in a mattress or in a cookie jar or behind an encyclopedia is a way of locking it up from burglars, though not a very safe one.

Although similar descriptions have been used by other authors and journalists, they show the writer's ignorance of locks. According to such a description, virtually anything even remotely involved in restricting access—from a door to a security guard—could be considered a lock. (That makes as much sense as saying a person's pants pocket is a safe or vault.) If locksmiths used such a broad definition, they wouldn't be able to talk about locks in a meaningful way.

For locksmithing purposes, a *lock* is a device that incorporates a bolt, cam, shackle, or switch to secure an object, such as a door, drawer, or machine, to a closed, locked, on, or off position and that provides a restricted means of releasing the object from that position. Although most locks are mechanical and are operated with a key, some are electric or computerized and are operated by pushing buttons, flipping an electric switch, inserting a specially coded card in a card reader, or providing biometric input.

Lock Names

Most locks have several names, which are usually based on a lock's common uses, appearances, major security features, installation method, internal construction, technology, or manufacturer. Many of the names have overlapping meanings. Some names based on common usage, like *trailer lock* and *gun lock*, are specific enough to be used by locksmiths as well as laypersons (Figures 3.1 and 3.2) because there are few significant variations among such locks. That is to say, one gun lock isn't much different from another gun lock.

Other names based on common usage, however, like *house lock* and *car lock*, refer to too many different types of locks. A locksmith would be confused if, say, a homeowner simply asked for a new "house lock." The homeowner would need to be more specific. He or she would need to ask for a key-in-knob or lever lock (names based on the style of handle a lock uses) (Figures 3.3 and 3.4). Other common "house locks" might include an *interlocking deadlock* or *deadbolt lock,* whose names are based on important security features (Figures 3.5 and 3.6).

While talking with one another about servicing locks, locksmiths often use names that describe a lock's installation method or internal construction, such as rim, mortise, and bored. A *rim lock* is any lock designed to be mounted on the surface (or rim) of a door or object. The interlocking deadbolt shown in Figure 3.6 is one type of rim lock. A *mortise lock* is installed in a hollowed out (or mortised) cavity (Figure 3.7). A *bored (*or *bored-in) lock* is installed by cross-boring two holes—one for the cylinder and one for the bolt mechanism. Warded, lever tumbler, disc tumbler, and pin tumbler are names that describe a lock's internal construction.

Figure 3.1 A trailer lock guards against tow-away theft of unattended boat, camping, snowmobile, horse, mobile home, and utility trailers. (*Master Lock Company*)

Figure 3.2 A gun lock protects against tampering. (*Master Lock Company*)

Figure 3.3 A key-in-knob lock is one of the most common locks used on homes. (*Master Lock Company*)

Figure 3.4 A lever lock is basically a key-in-knob lock with levers rather than knobs. (*Master Lock Company*)

Figure 3.5 A typical interlocking deadlock. (*Rathjen*, Locksmithing: From Apprentice to Master, *1995 McGraw-Hill, Inc., reproduced with permission*)

Deadbolt

Thumb turnpiece

Vertical

Deadbolt lever

Figure 3.6 A deadbolt lock is one of the most secure types of locks commonly found on homes. (*Arrow Lock Company*)

Figure 3.7 Most of a mortise lock is installed in a cavity or cutout. (*Arrow Lock Company*)

Figure 3.8 Warded bit key locks come in rim and mortised styles. (*Taylor Lock Company*)

Figure 3.9 Warded padlocks are for low-security applications. (*Master Lock Company*)

Warded locks

A *ward* is a fixed projection designed to obstruct unauthorized keys attempting to enter or operating the lock. One old type of warded lock comes in a metal case, has a large keyhole, and is operated with a bit key (commonly called a *skeleton key*). Such locks, called *bit-key locks*, come in mortised and surface-mounted styles and are often used on closet doors and cabinets (Figure 3.8). Some low-cost padlocks are also warded. Such a padlock can be identified by its wide, sawtoothlike keyway and keys with squared cuts (Figure 3.9). Warded locks provide little security because wards are easy to bypass with a stiff piece of wire or thin strip of metal.

Tumbler locks

Tumblers are small objects, usually made of metal, that move within a lock cylinder in ways that obstruct a lock's operation until an authorized key or combination moves them into alignment. There are several types of tumblers; they come in a variety of shapes and sizes, and move in different ways. Because tumblers generally provide more security than wards, most locks today use some type of tumbler arrangement either instead of or in addition to wards.

A typical key-operated cylinder consists of a cylinder case (or housing), a plug (the part with a keyway), springs, and tumblers. The springs are positioned in a way that makes them apply pressure to the tumblers. The tumblers are positioned so that when no key is inserted, or when the wrong key is inserted, the spring pressure forces one or more of the tumblers into a position that blocks the plug from being rotated. When the proper key is inserted into the keyway, however, the key moves the tumblers to a position that frees the plug to turn.

A lock can have more than one cylinder. A key-operated, single-cylinder lock has a cylinder on one side of the door only (usually the exterior side) so that no key is needed to operate it from the other side. Typically, it can be operated from the noncylinder side by pushing a button or by turning a knob, handle, or turn piece. Key-operated, double-cylinder locks require a key on both sides of a door. Many local building and fire codes restrict the use of double-cylinder locks on doors leading to the outside because the locks can make it hard for people to exit quickly during a fire or other emergency.

Types of tumbler locks.   There are three basic types of tumblers: lever, disc, and pin. Most *lever tumbler locks*, such as those used on luggage, brief cases, private mailboxes, and lockers, offer a low level of security. The lever tumbler locks commonly found on bank safe-deposit boxes are specially designed to offer a high level of security. *Disc tumbler locks* offer a medium level of security. They're often used on desks, file cabinets, and automobile doors and glove compartments. *Pin tumbler locks* can provide medium to high security, but in general they offer more security than do other types of tumbler locks. Many prison locks and virtually all house locks and high-security padlocks use pin tumbler cylinders. Some automobile door and ignition locks also have pin tumblers.

A special type of pin tumbler lock, called a *tubular key lock* (or tubular lock), has its tumblers arranged in a circular keyway (Figure 3.10). It uses a tubular key

Figure 3.10 A cross-section of a tubular key lock. (*Desert Publications*)

Figure 3.11 Types of IC locks. (*Arrow Lock Company*)

Figure 3.12 An IC core can be inserted into many different IC locks. (*Arrow Lock Company*)

to push the tumblers into proper alignment. Because of its odd appearance, a tubular key lock is harder to pick open than are most standard pin tumbler locks. Sometimes erroneously called "Ace Locks" (which is Chicago Lock Company's trade name for some of its tubular key locks), tubular key locks are often found on vending machines, laundromat equipment, bicycle locks, and high-security padlocks.

Another type of pin tumbler system is found in *interchangeable core* (IC) locks. Although they come in the form of deadbolts, key-in-knobs, rim locks, mortise locks, padlocks, and desk and cabinet locks, all locks in an IC system can either use the same key or be masterkeyed. Some examples of IC locks are shown in Figure 3.11. The common feature of IC locks is a figure 8-shaped core that houses the tumblers and springs. The cores can be easily removed and replaced (Figure 3.12). Any IC lock can be rekeyed simply by inserting new cores in the locks.

Other types of locks

Combination locks are popular alternatives to key-operated models. The two basic styles are pushbutton and dial. *Pushbutton* combination locks are operated by pushing a specific sequence of buttons, which are usually labeled with letters or numbers (Figure 3.13). *Dial* combination locks are operated by rotating one or more dials to specific positions (Figure 3.14).

An *electrical lock* can be operated by electric current. One type, sometimes called an *electric lock*, is basically a bolt or bar mechanism that doesn't have a keyed cylinder, knob, or turn piece and can't be operated mechanically. Another type, called an *electrified lock,* is a modified mechanical lock that can be operated either mechanically or with electricity. *Electric switch locks* complete and break an electric current when an authorized key is inserted and turned. An automobile ignition lock is an example of such a lock; after the key is turned, electricity flows from the battery to the car's starter. Similar locks are also used on alarm system control boxes to arm and disarm the system.

Time locks are designed to be opened only at certain times on certain days. They're commonly installed inside bank vaults and safes. *Biometric locks* unlock only after a computer has verified a physical feature, such as a fingerprint, signature, voiceprint, hand geometry, or the pattern of the retina of the eye.

Figure 3.13 Some combination locks use pushbuttons. (*Simplex Access Controls Corp.*)

Figure 3.14 Many padlocks have dial combinations. (*Master Lock Company*)

Lock Grades

The American National Standards Institute, Inc. (ANSI) determines manufacturing standards for a wide variety of building hardware and other products used in the United States. Many manufacturers make sure their products meet or exceed ANSI standards because architects, home builders, locksmiths, and other professionals specify products based on the standards.

As a rule, ANSI doesn't create standards. Interested industry associations usually create and propose standards and ANSI reviews them for possible adoption. The current standards for door hardware, listed under ANSI section 156, were proposed by the Builders Hardware Manufacturers Association, Inc. (BHMA).

ANSI 156 is of special importance to locksmiths. That section includes standards for many products locksmiths sell and install, including the following:

butts and hinges	ANSI/BHMA A156.1
bored and preassembled locks and latches	ANSI/BHMA A156.2
exit devices	ANSI/BHMA A156.3
door closers	ANSI/BHMA A156.4
auxiliary locks and associated products	ANSI/BHMA A156.5
architectural door trim	ANSI/BHMA A156.6
template hinge dimensions	ANSI/BHMA A156.8
cabinet hardware	ANSI/BHMA A156.9
power pedestrian doors	ANSI/BHMA A156.10
cabinet locks	ANSI/BHMA A156.11
interconnected locks	ANSI/BHMA A156.12
mortise locks and latches	ANSI/BHMA A156.13
sliding and folding door hardware	ANSI/BHMA A156.14
closer holder release devices	ANSI/BHMA A156.15
auxiliary hardware	ANSI/BHMA A156.16
hinges and pivots	ANSI/BHMA A156.19
strap and tee hinges and hasps	ANSI/BHMA A156.20
thresholds	ANSI/BHMA A156.21
electromagnetic locks	ANSI/BHMA A156.23
delayed egress locks	ANSI/BHMA A156.24

ANSI 156.2 includes provisions for locks to be Grade Certified. For a model to receive a grade, which ranges from 1 to 3, a sample lock must pass many rigorous tests that examine the lock's performance, strength, and durability. Grade 1 locks are the strongest and are often specified for industrial applications. Grade 2 locks are for light commercial and residential uses. Grade 3 locks are the lightest and are primar-

ily for residential applications. Some of the things the grading tests measure include bolt strength, turning torque needed to retract the latch, how well the finish holds up against salt sprays and other corrosives, and how many times the lock can be operated before failure.

Operational test

The operational test examines the amount of torque needed to retract the latchbolt with and without a key. It's performed first by depressing the deadlatch plunger (if necessary), then slowly applying torque force to the outside knob of an unlocked lockset until the latch is fully retracted. For a key-in-knob to receive a grade, torque used may not exceed 9 lbf-in. For a lever lock to be graded, torque cannot exceed 28 lbf-in. (Key-in-knobs are then tested in the opposite direction.) The test is repeated for the inside knob or lever.

The lock is then put in the locked position and the deadlatch plunger is again depressed (if applicable). Then the key is inserted into the keyway and is slowly rotated until the latch is fully retracted. Torque may not exceed 9 lbf-in. The test is repeated in the opposite direction if the lockset is designed to allow such movement. If the inside knob or lever is key-operated, the test is also applied to the inside knob or lever.

Strength test

The strength test examines how much forcible turning force a lock in the locked position can withstand. To be graded, a lock must stay locked after the minimum force has been applied. For a Grade 1, a key-in-knob must hold up to 300 lbf-in.; a lever lock must withstand 450 lbf-in. For a Grade 2, a key-in-knob must hold up to 150 lbf-in.; a lever lock must hold up to 225 lbf-in. A Grade 3 requires a key-in-knob to withstand at least 120 lbf-in. and a lever lock to withstand at least 180 lbf-in.

Cycle test

The cycle test examines how many times the lock can be operated before failure. For a Grade 1, a lock must complete 800,000 cycles. For a Grade 2, a lock must complete 400,000 cycles. Grade 3 locks complete at least 200,000 cycles.

Key Types

There are six basic types of keys: bit, barrel, flat, corrugated, tubular lock, and cylinder. Although different types of keys have different parts, virtually all keys have a bow (rhymes with toe). The bow (or handle) is the part of the key that a person holds while inserting the key into a lock. Bows come in a variety of shapes and often have identifying information or advertising imprinted on them. Figure 3.15 shows some common types of keys.

A *bit* (or skeleton) key is usually made of iron, brass, steel, or aluminum. Major parts of the key include the bow, shank, shoulder, post, and bit. Many *barrel* keys look similar to bit keys, but barrel keys have a hollow shank and often don't have a shoulder.

Figure 3.15 Different types of keys have different parts.

A *flat* key is flat on both sides. Most are made of steel or nickel silver. Such keys are used for operating lever tumbler locks. *Corrugated* keys look similar to flat keys. Both types usually have the same parts. However, corrugated keys have corrugations (ripples) along the length of their blades. Corrugated keys are most often used to operate warded padlocks.

A *tubular* key has a short tubular blade with cuts (depressions) milled in a circle around the end of the blade. The key is used to operate tubular key locks. Parts of a tubular key include the bow, blade, tumbler cuts, and nib. The *nib* is a small protrusion at the tip. It shows the position the key must be at to enter and operate the lock.

The most common type of key is the *cylinder* key. It's used for operating most pin tumbler locks and disc tumbler locks. Major parts of a cylinder key include the bow, shoulder (one or two), blade, keyway (or milling) grooves, and tip. The shoulders of a key are generally used as stops when cutting keys. When a cylinder key doesn't have a shoulder, its tip is used as the stop. The keyway grooves are millings along the length of the blade that allow the key to fit into a lock's keyway.

Key Blank Identification

It's as important for locksmiths to be able to identify keys and key blanks as it is for them to identify locks. A *key blank* is basically an uncut key. Before duplicating a key, a locksmith must find a matching blank.

Important factors to consider when choosing a blank to duplicate a bit key or barrel key are thickness of the bit and diameter of the shank and post. Because there's usually a lot of tolerance in locks that use these types of keys, the blank might not have to match perfectly.

Important factors when choosing a blank for flat keys and corrugated keys are thickness, length, width, and shape of blade. The key and blank should match closely in those areas. A blank for a corrugated key should have the same corrugated configuration that the key has.

Choosing a blank for a tubular key is simple because there are few significant differences among tubular keys. The important areas of such keys are the size of the nib and the inside and outside diameters of the shank. If you find a blank that closely matches the key in those respects, you can duplicate the key.

Cylinder key blank identification

The trickiest type of key to find a blank for is the cylinder key, the type that locksmiths are most often asked to duplicate. Three parts that are generally common to cylinder keys and their corresponding key blanks can be helpful when searching for a key blank. These are the bow, blade length, and keyway grooves.

The *bow* is the head of the key or blank. It's the part that you grip when using a key to operate a lock. Most cylinder lock manufacturers use distinctive key bows and many aftermarket key blanks copy their basic shapes. Based on the bow alone, a locksmith can often either quickly find a matching key blank or at least narrow the choices to a few key blanks.

The *blade* is the part of the key that enters a lock's keyway. Generally, the key and key blank should have blades of the same length. To compare lengths, you can either hold the key and blank together or use the illustrations in a key blank catalog. When using a catalog, lay the key over the illustration and align the shoulders. You can then observe the length at a glance because the key catalog drawings are the size of the original keys.

Keyway grooves (or millings) are critical. Only when the key and key blank have the same, or very similar, grooves can both fit into the same keyway. One way to compare millings is to try to insert the key blank into the lock's keyway. If it can be inserted, the millings are the same or similar. You can also compare millings by holding the key and key blank side by side and looking at the blade tips. Finally, use the catalog to compare blade millings. To use the catalog, stand the key directly over the cross-section that appears under the blank's illustration. The grooves must match exactly to be considered the same.

Key manufacturers publish catalogs that identify their key blanks and sizes and show a cross-section of each different blank. In addition, they may have a cross-reference section which refers to other manufacturers' keys that are comparable to each of the keys in that particular catalog. These cross-reference sections are valuable because they obviate reference to a wide variety of catalogs in order to determine the manufacturer of a given key blank. Merely refer to the key type and cross-reference it to the one that you have.

Some manufacturers have literally hundreds of key blanks and the cross-reference section becomes unwieldy at times. It is best to use a catalog that carries the most popular and commonly used blanks. About 95 percent of the keys you will duplicate will be in such catalogs. Other catalogs can be used for general reference but are not absolutely necessary to have for your shop.

Key Blank Examples

On the following pages are illustrations of various types of cylinder key blanks. Figure 3.16 shows key blanks from the Taylor Lock Company. These blanks will fit a variety of American locks. Notice that in Figure 3.17 the blank is further laid out to incorporate a master key system and that, from the cross-section, you can identify the various keyways that will fall into this particular master key system.

Figure 3.18 shows another keyway system with the applicable blanks. Notice that within this key series there are three different key lengths, which allow for wider use of the master key system. The length also allows more pins to be used in the cylinders. This means a greater variety in the number of possible key combinations available to the system.

Foreign automobiles (and, naturally, their keys) are increasing in popularity, so it is necessary to have a well-stocked variety of foreign lock keys in the locksmith shop. Figure 3.19 provides illustrative details on a variety of the more common key blanks in use. You should have these types and a slightly wider supply of other foreign auto key blanks available to service the needs of your customers.

Figure 3.16 Representative sampling of standard key blanks. (*Taylor Lock Company*)

Figure 3.17 Six-pin cylinder key and various associated keyways. Keyway M is used only for the master key. (*Taylor Lock Company*)

LN = 49 KM (5 pin)
049 KM (6 pin)
A49 KM (7 pin)

To order keys, keyway section is added to pin length. Example - 43 LA is 5 pin length, having LA keyway and A43LH is submaster in 7 pin length to by pass LE, LF and LG key sections.

43 Series
5 pin

O43 Series
6 pin

A43 Series
7 pin

Figure 3.18 A large master key system provides a great variety of key uses and control features through sub-masters and individual use of keys. (*Taylor Lock Company*)

Automotive Key Blanks

Duplicating automotive keys is more prevalent in recreational areas and very large cities. Lost or stolen keys mean duplication and/or rekeying on a daily basis for many locksmiths. It is imperative that, as a locksmith, you have in your shop an ample selection of automotive key blanks on hand. Keep a selection of the most common and popular automotive blanks in your van plus one or two others that might be required. (*Note:* Every locksmith shop should have a standard selection of automotive key blanks, but in large cities and in recreational areas such as beaches, parks, etc., the need is greater than usual. Thus, the number and variety of blanks needed will be greater in these areas than in others. Always consult with your distributor or manufacturer's representative to obtain an accurate view of the types and quantities of blanks that are necessary; don't buy a larger selection than required just because you think you may someday need it.)

Figure 3.20 is a list of foreign automotive blanks identified by the number and the vehicle they accompany. These are, by far, the most common vehicles for which you

Types of Locks and Keys 45

will be cutting keys. (This is a representational selection of popular keys that require frequent duplication; it is not all-inclusive.) Refer to your key catalog for illustrations of all the automotive keys and develop a working memory of them. After a period of

Figure 3.19 Key blanks for foreign automobiles. (*Taylor Lock Company*)

AD1	Audi 100 and 100LS—1971 on
CP1	Capri Ign. 1971 on—Ign./Trunk 1975 on
CP2	Capri Door/Trunk 1971-74
CP3	Capri—Supplements CP2
CP4	Capri—Fiesta
5DA1	Datsun
5DA2	Datsun 1970-72, Mazda 1971 on, Eng. 1969 on
DA3	Datsun—1970 on
DA4	Datsun & Subaru—1970 on
LDC1	Long Head Dodge Colt, Arrow, Chall., Sapporo, Jap. Opel—1970 on
5FT1	Fiat Ignition—1967 on
FT2	Fiat Trunk 1967 on—Pántera-Ferrari
FT3	Fiat—Supplements FT2
HN1	Honda Ignition 1970 on, MG Ignition 1972
HN2-3	Honda MASTER (Codes #1001-1700, 2001-2700)—1976
HN4	Honda Trunk 1973-76
HN5	Honda Cycles—1977 on

LU1	GM LUV Ign./Door & Trunk/Glove—1973 on
MZ1	Mazda Door/Trunk—1971 on
MZ2	Mazda—Supplements MZ1
MZ3	Mazda, Courier Trucks 1971-73
MZ4	Mazda/Courier R100/1200-Door/Trunk 1970-72
MZ5	Mazda—Supplements MZ4
RP1	Renault—Peugeot
5RP2	Renault-Peugeot Ignition 1971 on
TO1	Toyota
TO2	Toyota Ignition—1969 on
UN2	Union (Hillman—Vauxhall—English Ford)
UN3	English 1960 on, Volvo Door/Trunk 1970
UN4	Union—MG—Triumph
6VL1	Volvo 1960 to 1965-6 Pin
VW1	Volkswagen
VW2	Volkswagen 1965-70
VW3	Volkswagen Beetle—1971 on
VW4	Volkswagen VW411-412-DA-RA-SC, Fox, Audi 4000/5000, Porsche, Jetta
VW5	Volkswagen Bus—1971 on

Figure 3.20 Foreign automobile key blanks. (*Star Key and Lock Co.*)

Figure 3.21 Sampling of common domestic automobile key blanks. (*Star Key and Lock Co.*)

time you should be able to look at a key that is already cut (or even a blank) and determine with relative accuracy its make of automobile and where the key blanks are located on your key blank board. Figure 3.21 shows some domestic key blanks. Figure 3.22 lists the most common domestic keys by number of automotive type and Figure 3.23 shows a variety of foreign key blanks.

Look-Alike Keys

You will find key blanks that are *look-alike* keys—blanks made by a manufacturer other than the original with bows that look like the bows of the original key blanks. These provide you with a great advantage; as look-alikes, they are quickly identified with a specific lock manufacturer, enabling you to quickly select the proper key for duplication.

Because not all keys are look-alikes you will need comparison or cross-reference key charts—important reference tools. Figure 3.24 is a manufacturers' key comparison chart for automotive and house keys.

Suppose you have a customer who brings in two keys for duplication—one for her house and one for her car. The house key is National/Curtis number IN1 and the car key is 1127DP. As an up-and-coming locksmith, you carry a wide assortment of the most commonly used keys and you have a cross-reference key chart listing. You can quickly determine whether or not you have the particular blanks for these two keys.

Go to the cross-reference list and find that the IN1 goes to the Star blank 5IL1 (Ilco house lock). The 1127DP is an Ilco key for which you have the Star blank,

Code	Description
HFD1 / OFD1	Ford Ignition/Door—1952 to 1954 / Ford Trunk/Glove—1952 to 1954
HFD2 / OFD2	Ford Ignition/Door—to 1951 / Ford Trunk/Glove—to 1951
HFD3 / OFD3	Ford Ignition/Door—1955 to 1958 / Ford Trunk/Glove—1955 to 1958
HFD4 / OFD4	Ford Ignition/Door—1952 to 1964 / Ford Trunk/Glove—1952 to 1964
HFD5 / OFD5	Mercury Ignition/Door—1952 to 1964 / Mercury Trunk/Glove—1952 to 1964
HFD6	Ford, Linc., Merc. Ign./Trunk Square Head Master—1952 to 1964
OFD6	Ford, Linc., Merc. Ign./Trunk Round Head Master—1952 to 1964
HFD7	Ford Ign./Door to 1951—Studebaker 1947 on
HFD8	Lincoln Ignition/Door—1952 to 1964
HFD9 / OFD9	Ford Ignition/Door—1965/66 only (double side) / Ford Trunk/Glove—1965/66 only (double side)
HFD10 / OFD10	Ford Ignition/Door—1965 on (double side) / Ford Trunk/Glove—1965 on (double side)
HPL1 / OPL1	Chrysler, Plymouth, Dodge Ign./Door 1949-55 / Chrysler, Plymouth, Dodge Trunk/Glove 1949-58
HPL2 / OPL2	Chrysler, Plymouth, Dodge Ign./Door to 1948 / Chrysler, Plymouth, Dodge Trunk/Glove to 1948
HPL3	Chrysler, Plymouth, Dodge Ign./Door 1956-59
OPL4 / OPL5	Chrysler, Dodge Trunk/Glove 1959-65 / Plymouth Trunk/Glove 1959-65
HPL6	Chrysler, Plymouth, Dodge Ign./Door 1960-67
OPL7	(Master) Long GM to 1966 & Chrysler Trunk 1959-65
OPL8	Chrysler Trunk/Glove—1966/67
HPL68 / OPL68	Chrysler Ign./Door Master 1956 on—hexagon head / Chrysler Trunk/Glove Master—1966 on
OPL70	Chrysler Ign./Door & Trunk/Glove Master—1966 on
HPL73	Chrysler Ign./Door Master 1956 on—diamond head
HYA4	Kaiser Frazer Ign./Door—1949/50
HYA5	Studebaker Ign./Door—1940 on

Code	Description
HBR1 / OBR1	American Motors Ignition/Door to 1959 / American Motors Trunk/Glove to 1959, GAS locks
OBR1DB	Briggs & Stratton—GAS & utility locks
HBR2 / OBR2	GM Ignition/Door—all years to 1966 / GM Trunk/Glove—all years to 1966
HBR3 / OBR3	American Motors Ignition/Door—1960 on / American Motors Trunk/Glove—1960 on
OBR4	GM Long Head Ignition/Door—all years to 1966
HBR5 / HBR5M / OBR5	GM Ignition/Door "A"—1967 / GM Ignition/Door Master "A/C"—1967/68 / GM Trunk/Glove "B"—1967
HBR6	GM Ign./Door & Trunk/Glove Master—hexagon head—1967
OBR6	GM Ign./Door & Trunk/Glove Master—oval head—1967
HBR7 / OBR7	GM Ignition/Door "C"—1968 / GM Trunk/Glove "D"—1968
HBR8	GM Long Head Ign./Door Master "A/C"—1967/68/71/72 (Ign. only 1975/76/79/80)
HBR9	GM Long Head Ign./Door & Trunk/Glove Master "E/H"—1969/73/77/81
OBR9	GM Oval Head Ign./Door & Trunk/Glove Master "E/H"—1969/73/77/81
HBR9E / OBR9H	GM Ign./Door "E"—1969/73 (Ign. only—1977/81) / GM Trunk/Glove "H"—1969/73 (Door also—1977/81)
HBR10J / OBR10K	GM Ign./Door "J"—1970 (Ign. only—1974/78/82) / GM Trunk/Glove "K"—1970 (Door also—1974/78/82)
HBR11	American Motors Ign./Door—1970 square head—1960 on
OBR11	American Motors Trunk/Glove—1970 ellipse head—1960 on
HBR12A / OBR12B / OBR13	GM Ign./Door "A"—1967/71 (Ign. only—1975/79) / GM Trunk/Glove "B"—1967/71 (Door also—1975/79) / GM Ign./Door & Trunk/Glove Master "A/B"—1967/71/75/79
OBR13S	GM Ign./Trunk—fits most years—1967 on—FOR LOCKSMITHS ONLY—NOT FOR DUPLICATION
HBR14C / OBR14D	GM Ign./Door "C"—1968/72 (Ign. only—1976/80) / GM Trunk/Glove "D"—1968/72 (Door also—1976/80)

Figure 3.22 Domestic automobile key blanks. (Star Key and Lock Co.)

Figure 3.23 Sampling of common foreign automobile key blanks. (*Star Key and Lock Co.*)

HFD4—a Ford auto ignition key. Rapidly looking at the key blanks, you quickly se-lect the two proper keys. Within minutes you have another satisfied customer, thanks to the cross-reference listing.

The cross-reference listing is available from your key blank representative, or you can spend between $20 and $50 to get a complete cross-reference book. This is both an advantage and a disadvantage. With the full book, you may have to look in a number of sections whereas with the smaller, individual cross-reference breakout from your key distributor/manufacturer, you are more likely to find the required key quickly.

A copy of a cross-reference listing is a prime requisite for every shop. Remember, your priority is the listing obtained from your key manufacturer; the book can come later if you find you need it.

Neuter Bows

Because most nonlocksmiths can only identify key blanks by their bows, locksmiths often use *neuter* (or security) bow key blanks to make keys harder to duplicate. Such bows have a generic shape and style and provide no information that identifies

AUTOMOTIVE

STAR	Ilco	Natl/Curtis	Make
HBR2	H1098LA	B10	GM Ignition to 1966
OBR2	01098LA	B11	GM Trunk to 1966
HBR3	H1098NR	B24,RA2	Amer. Mtrs. Ign. 1960 on
OBR3	1098NR	B4,RA1	Amer. Mtrs. Trunk 1960 on
HBR5	H1098A	B40	GM Ignition "A"-1967
HBR9E	P1098E	B44	GM Ignition "E" - 1969 & 1973 (Ign. only-1977/81)
OBR9H	S1098H	B45	GM Trunk "H" - 1969 & 1973 (Door also-1977/81)
HBR10J	P1098J	B46	GM Ignition "J" - 1970 (Ignition only-1974/78/82)
OBR10K	S1098K	B47	GM Trunk "K" - 1970 (Door also-1974/78/82)
HBR11	1970AM	RA4	Amer. Mtrs. Ignition - 1970 square head - 1960 on
OBR11	S1970AM	RA3	Amer. Mtrs. Trunk - 1970 ellipse head - 1960 on
HBR12A	P1098A	B48	GM Ignition "A" - 1971 & 1967 (Ign. only-1975/79)
OBR12B	S1098B	B49	GM Trunk "B"-1971 & 1967 (Door also-1975/79)
HBR14C	P1098C	B50	GM Ignition "C" - 1968/72 (Ign. only-1976/80)
OBR14D	S1098D	B51	GM Trunk "D" - 1968/72 (Door also-1976/80)
HFD4	1127DP	H27	Ford Ignition 1952-64
OFD4	1127ES	H26	Ford Trunk 1952-64
HFD10	1167FD	H33,H51	Ford Ign.Double Side 1965 on
OFD10	S1167FD	H32,H50	Ford Trk.Double Side 1965 on
OPL4	1759L/1764S	Y141,B29-31	Chrysler Trunk 1959-65
HPL68	1768/69/70CH	Y152/150/148/146	Chrysler Ign.Master 1956 on
OPL68	S1768/69/70CH	Y151/149/138	Chrysler Trk.Master 1966 on

HOUSE

STAR	Ilco	Natl/Curtis	Make
5CO1	1001EN	CO7	Corbin X1-67-5
5DE3	D1054K	DE6	Dexter 67
5EA1	1014F,X1014F	EA27,EA50	Eagle 11929 - Harloc
5HO1	1170B	HO3/HO1	Hollymade - Challenger K1
5IL1	1054F	IN1	Ilco 1054F - Keil 159AA
5IL2	1054K	IN3	Ilco - Dexter - Weslock
5IL7	X1054F	IN21,IN18	Ilco X1054F
5KE1	1079B	K2	Keil 2KK
5KW1	1176	KW1,PT1	Kwikset 1-1063 - Donner - Petco
5LO1	1004	L1	Lockwood B308 - 5 Pin
MA1	1092	M1	Master 1K,77K
5RO4	R1064D	NA5, NA6	Rockford 411-31A
5RU2	1011P	RU4	Russwin 981B
5SA1	01010	S4	Sargent 265U
5SE1	1022	SE1	Segal K9 - Norwalk - Star
5SH1	1145	SC1	Schlage 35-100C(923C)
SH2	1307A	SC6	Schlage 35-180(920A)
SH6	1307W	SC22	Schlage 35-200(927W)
5TA4	1141GE	T4,T7	Taylor 111GE
5WK1	1175, 1175N	WK1/WK2	Weslock 4425
5WR2	1054WB	WR2,WR3	Weiser 1556 - Falcon
4YA1	999B	Y145,Y220	Yale 9½
5YA1	999	Y1	Yale 8 - Acrolock

Figure 3.24 Key comparison chart for automotive and residential key blanks. (*Star Key and Lock Co.*)

a lockmaker. That prevents unauthorized people who may have a key for a short time from running to any hardware or department store and getting the key copied. In addition to increasing security, such bows provide space to imprint advertising, increasing the likelihood that the customer will return to the same locksmith to get duplicate keys. Figure 3.25 shows some neuter bow key blanks.

Embossed logo

Custom incised logo

Standard incised design

Figure 3.25   Neuter bow key blanks can be used for advertising purposes. (*Kustom Key, Inc.*)

Chapter

4

Warded Locks

The warded lock is the oldest lock still in use and is found in all corners of the world. It employs a single or multiple warding system. Because of its simple design, its straightforward internal structure, and its easily duplicated key, this lock is an excellent training aid for locksmiths. This same simplicity means that warded locks give very little security. Use these locks only in low-risk applications such as storage sheds and rooms where absolute security is not essential.

At one time warded locks were used on most doors. They are still found in abundance in buildings still standing in older metropolitan neighborhoods such as Center City Philadelphia, Market Street in San Francisco, the Old Town section of Chicago, and the East Side of New York City.

The oldest of these buildings have cast-iron locks on the doors, some of which date back to the last century. Later types were made of medium-gauge sheet metal. The casing consists of two stampings: the cover plate and the back plate. The latter mounts the internal mechanism and forms the sides.

The warded lock derives its name from the word *ward*, meaning to *guard*. The interior of the lock case has protruding ridges or wards that help protect against the use of an unauthorized or improperly cut key. Normally there are two interior wards positioned directly across from each other on the inside of the cover and backing plates.

This lock is sometimes mistakenly referred to as a *skeleton-key* lock. The proper and full name is the *warded-bit* key lock.

Types

Two types of warded locks are currently in use: the *surface-mounted* (or *rim*) lock and the *mortised* lock (Figure 4.1). While both types are similar in structure and size, they give a varying degree of security. The internal mechanisms of both operate

51

Figure 4.1 Bit key locks are available in mortised (left) and surface-mounted (right) styles. (*Taylor Lock Co.*)

on the same principle, but the mortise lock may have several additional parts. Differences between these locks are as follows:

Surface-mounted (rim) lock	Mortised ward lock
Mounted on door surface	Mounted inside of door
Secured by screws in the door face	Secured by screws in the side if the door at the lock face plate
Door can be any thickness	Door must be thick enough to accommodate
Thin case	Fairly thick case
Short latchbolt throw	Up to 1-inch latchbolt throw
Lock from either side	Locked from either side
Strike can be removed with door closed	Strike cannot be removed with door closed
Very restricted range of key	Restricted range of key changes
Very weak security	Weak security

Construction

The basic interior mechanism is drawn in Figure 4.2. Since the relative security of any lock lies in the type of key used, the number of key variations possible, and the amount of access to the locking mechanism afforded by the keyhole, the warded lock is the least secure.

The keyhole is an access route to the interior mechanism of the lock. The larger the keyhole, the easier it is to insert a pick or other tool and release the bolt.

If a lock was designed to have no more than 10 different key patterns (changes) and 1000 locks were made, 10 different keys would open all 1000 locks. By the same token, one key would open the lock it was sold with and 99 others. Furthermore, it is often possible to cut away parts of a key to pass (negotiate) the wards of all 1000 locks. You can see that lock security is related to the kind and number of key changes built into the system when it is initially designed.

In theory, each warded lock can be designed to accept 50, or even 100, slightly different keys. In practice, these locks tend to become more selective as they age and wear. The lock might respond to the original key or to one very much like it, but other keys that would have worked when the mechanism was new no longer fit. While this might seem fine and well for the lock owner, excessive wear increases the potential of both key breakage within the lock and jams in the open, partly open, or closed positions. It can also mean that a new lock will have to be installed.

Most surface-mounted and mortised locks are intended to be operated from both sides of the door. Keyholes and doorknob spindle holes extend through both sides of the lock body. Occasionally you will encounter a surface-mounted lock with a doorknob spindle and keyhole only on one side. The other side is blanked off. A lock of this type can be modified to accept a key from the other side. This modification entails cutting a keyhole through the door and lock body and may require some filing on the key. Figure 4.3 illustrates the differences in keys. Note the additional cut on the left-hand key.

Figure 4.2 The internal parts of a typical bit key lock.

WARD

DOOR
THICKNESS

Figure 4.3 The key on the right is for a lock activated from one side of a door; the key on the left can pass the wards from either side.

Operation

The key must be cut to correspond to the single or multiple side and end wards that have been designed into the lock. After the key passes these wards, it comes in contact with the locking mechanism. The cuts on the key lift the lever to the correct height and throw the deadbolt into the locked or unlocked position. Turning the doorknob activates the spindle and, so long as the deadbolt is retracted, releases the door.

Figure 4.4 depicts various keyhole control features that allow only certain types of cut keys to enter the keyhole. Figure 4.5 shows a key entering a keyhole. Notice that the key has the appropriate side groove to allow it to pass through the keyhole and into the lock. If you were to file off this obstruction (called a *case ward*), any key thin enough to pass could enter. (Some ward bit keys are quite thick.) By the same token, a very thin key can pass whether or not the side ward is present (Figure 4.6). The common skeleton key is a prime example of this; it is thin enough to pass most case wards but it will not necessarily open the lock.

Figure 4.7 shows a key engaging the bolt. While there is only one set of wards in this particular lock, the positioning of this ward gives more security than a lock with no ward or one with a ward that has been worn down to almost nothing.

Repair

Because it is usually cheaper to replace them, warded locks are not repaired to any great extent. You should, however, have a supply of spare parts for these locks.

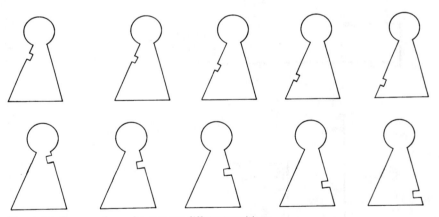

Figure 4.4 Case wards can be at many different positions.

Figure 4.5 A slot milled on the edge of the key allows the key to pass the case ward.

CUT AWAY

Figure 4.6 A skeleton key is a key that has been ground down to bypass the case ward.

LOCK BOLT

WARD

Figure 4.7 Even a single ward limits the number of keys that can operate a bit key lock.

Broken locks

The most frequent failure is a broken spring. Over a period of time, the spring may crystallize where it mates with the bolt. In addition, the wards can break or wear down into uselessness.

Replace a broken spring with a piece of spring stock cut to length and bent to the correct angle. Some spring stock must be tempered before use; other springs come already tempered. If tempering is necessary, heat the spring to cherry red then quench it in oil. You can save time by purchasing standard springs already bent into a variety of shapes that are designed to fit almost all locks.

Worn or broken wards on locks with cast cases can be repaired by drilling a small hole in the case and forcing a short brass pin into the lock case. The best technique for brittle cases is to braze a piece of metal on the case at the appropriate spot and file it down to the appropriate size. Wards on locks with sheet metal cases can be renewed by indenting the case with a punch ground to a fairly sharp point. If the factory has already punched out the wards, it would be best to braze a piece of metal at the proper spot.

Since most of these locks are inexpensive and offer minimal security, you should remind customers that the cost of repair may far exceed the cost of the lock. Purchasing a new and more secure lock has definite advantages. If you are already at his home on a call, the homeowner can save money by asking you to install a new lock immediately instead of ordering one and having you make a second trip to install it.

Should the homeowner decide to take your advice and purchase a new lock, ask to keep the broken one; it is of no use to her and parts are always nice to have. Sooner or later you will have to repair another lock of the same type; having the correct parts at hand will save you time. Furthermore, you have made a sale and, by being allowed to keep the old lock, have obtained parts at no cost.

Removing paint

Removing paint from warded locks is also a form of repair. The home painter is a major cause of lock failure. He often does not take the time to remove the lock or to cover it with masking tape. Paint usually gets into the mechanism, freezing its works. To clean a paint-bound lock, follow this procedure:

Remove the lock from the door. (Run a sharp knife around the edges so the new paint will not be cracked and broken.) Disassemble the lock. Using a wire brush, scrape the paint from the parts. In extreme cases, you may have to use a small knife or soak the individual parts in paint remover. Dry each part thoroughly. Check for rust and worn parts and replace as needed. Assemble and mount the lock.

Use paint remover only as a last resort because it leaves a residue that attracts dust and lint. When you use paint remover you must clean each part before assembly.

Lubricating locks

Locks that are difficult to operate usually have not been lubricated in a long time, if ever. Never use oil to lubricate a lock. The professional approach is to use a flake or powdered graphite. Apply the lubricant sparingly. Remember, a little bit goes a long way. This is especially true of graphite. Should you overuse it, you may have to explain to the homeowner why there is a dark patch on the carpet that cannot be cleaned. Graphite stains are almost impossible to remove.

Warded Keys

Warded keys are made of iron, steel, brass, and aluminum. Iron and aluminum keys have a tendency to break or bend within a relatively short time; steel and brass keys can outlast the lock. The warded key has seven parts, as shown in Figure 4.8. The configuration of the box, length of the shank, and the relative thickness of the shoulder are not critical to the selection or the cutting of the key.

BOW   SHANK   COLLAR   POST   THROAT   BIT   BITTING

Figure 4.8 Parts of a bit key.

Types of warded keys

Warded keys come in various types, including the simple warded key, the standard warded key, the multicut key, and the antique key. *Simple* warded keys are often factory-made precut keys that fit several different keyholes. *Multicut* keys, on the other hand, are designed for specific locks. The *standard* warded key is usually mass-produced, but it has more precut ward and end cuts than the simple warded key. Standard warded keys can be easily converted into master keys by cutting. The *antique* key may have several kinds of cuts: ward cuts, end cuts, and even side (or bullet) groove cuts extending the length of the bit. Antique keys usually go to older locks, but these keys are still manufactured.

Selection of key blanks

A key blank is a key that has not been cut or shaped to fit a specific locking mechanism. When selecting a blank for a duplicate warded key, the following should be considered:

Post size. The post of both keys (original and duplicate) must be the same diameter. If your eyesight cannot correctly determine whether or not they are the same size, use either calipers or a paper clip. Use the calipers to compare the diameter of the original key with the diameter of the duplicate. You can use a paper clip that has been wrapped around the original key to check the diameter of the duplicate.

Length. From the collar to the end of the pin, both keys should be approximately the same length. This is important because this portion of the key enters the keyhole and operates the lock.

Height. The height of the bitting (cuts in the bit) must be the same on both keys. If the bitting in the duplicate is higher, it should be filed down; if it is lower, another blank should be selected.

Bow. The bows need not be identical but generally should be closely matched.

Width. The width of the bittings should be approximately the same. If the bitting on the duplicate is too wide, the extra thickness may prevent the duplicate from entering the lock.

Thickness. The thickness of the bits should be the same. If the original key bit is tapered, the bit of the duplicate should also be tapered. You may have to select a blank with a thick bit that you can file down to the correct taper.

TABLE 4.1 Standard Wire Gauges (inches and millimeters). 1 mm = 0.03937 inches; 1 inch = 25.4 mm

Standard Wire Gauge Number	Inches	Millimeters	Standard Wire Gauge Number	Inches	Millimeters
4/0	0.400	10.16	7	0.176	4.47
3/0	0.372	9.45	7½	0.168	4.27
2/0	0.348	8.84	8	0.160	4.06
0	0.324	8.23	9	0.144	3.66
1	0.300	7.60	10	0.128	3.25
2	0.276	7.01	11	0.116	2.95
2½	0.264	6.71	12	0.104	2.64
3	0.252	6.40	13	0.092	2.34
3½	0.242	6.15	14	0.080	2.03
4	0.232	5.89	15	0.072	1.83
4½	0.222	5.64	16	0.064	1.63
5	0.212	5.38	17	0.056	1.42
5½	0.202	5.13	18	0.048	1.22
6	0.192	4.88	19	0.040	1.02
6½	0.184	4.67	20	0.036	0.91

If you don't have a micrometer, Table 4.1, which shows standard wire diameters, can help you determine the approximate diameter of warded key blanks. You can take a standard piece of wire with a known diameter and compare it with any key blank. Also, use a drill to determine the approximate diameter of warded key blanks. Insert the pin of a blank into the hole that matches the blank's diameter.

Duplicating a warded key by hand

1. Select the proper key blank.
2. Wrap a strip of aluminum approximately 1½ inches × 2¼ inches or 2½ inches around the pin and bit of the original key with one edge against the collar (Figure 4.9).
3. Clamp the original key (wrapped in the aluminum) into a vise, bitting edge up. Ensure that the aluminum fits snugly around the key bit and pin.
4. Cut off excess aluminum around the bit of the key. Remove the aluminum strip and smoke the key with a candle.
5. Place the strip back on the bit and reclamp.
6. Using a warding file, cut down the aluminum strip—*not the original key*—until it is in the shape of the original key. Since the aluminum bends easily, use the file only in one direction—away from you. The stroke should be firm and steady at

first. As you file closer to the cuts of the original key, the strokes should be shorter and lighter. When the file just barely touches the original key and starts to remove the candle black, *stop and go no further*. If the candle black is removed and the shiny surface of the original key is revealed, you have filed too deeply.

7. Fit the aluminum strip onto the key blank.

8. File down the exposed areas on the bit until it matches the outline of the aluminum strip. Be careful not to cut into the strip. Again, use shorter and lighter strokes as you get closer to finishing each cut.

9. If the original key has a side groove that matches a keyhole ward, this too must be cut. Use another strip of aluminum. If the original key has two grooves, you must wrap the strip around the bit so that both grooves are covered.

10. Using a scriber, scratch the metal strip to indicate the top and bottom of the groove(s). Fit the strip onto the key blank and mark the positions of the groove(s) on both ends of the bit. By connecting the marks with lines, you know exactly where to file.

11. To determine the depth of a groove cut, put one edge of a metal strip into the groove of the original key and scribe the depth of the groove on it. This mark will be your depth guide when filing the groove on the duplicate key.

The impression method

The term *impression* refers to a duplicate key that is made from the lock itself as opposed to copying the original key. Impressioning is a skill that sets the locksmith apart from the individual who merely duplicates keys in the local five-and-dime store. This skill is an extremely important and valuable asset.

Impressioning is the ability to decipher small marks made on a smoked key blank that has been inserted into a lock and turned. Interpretation of the marks tells the locksmith what cuts to make, where to make them, and how deep they must be. The advantage of impressioning is that you need not disassemble a lock or remove it from a door to make a key.

The first cut allows that blank to enter the keyhole. Refer to the previous section for information on what is required to make such a cut. Since you do not have the original key, smoke the end edge of the blank and insert it into the keyhole so the edge comes into contact with the case ward. Scribe mark the top and bottom of the ward

Figure 4.9 File into the aluminum strip in the direction shown.

on the blank. Remove the key; the candle black that was removed indicates the depth of the cut and the scribed marks show the position of the cut on the blank. Transfer the scribe marks to the near end of the bit and draw lines connecting the two pairs of marks. As in the previous section, use a small piece of metal to make a depth gauge so you will not file too deeply. Mount the blank in the vise and cut the ward slot. When you're finished, the key should pass the case ward.

Next, prepare the blank for impressioning the internal wards. Recall that the key must pass certain side wards. When the lock is assembled, how can you be sure exactly where to cut the key so it will pass these wards? You can't. You will have to prepare the key blank using one of two methods. The first is to coat the key with a thin layer of wax. This is unprofessional and can harm the lock. The wax may clog the mechanism, forcing you to disassemble and clean the mechanism *at your expense*— you can't charge a customer for your mistake. The professional method is to smoke the key. The smoking must be thick enough to form a stable marking surface. If, for example, the blank is thicker than it should be, enough blackening must be present to give true readings. A thin coat will speckle as you turn the key and send you on wild goose chases.

The technique for impressioning the key is as follows:

1. Insert the key into the lock and turn it with authority. Remove the key. You will notice one or more bright marks where the blackening has been removed. These marks indicate obstructions. File the appropriate cut for the key to pass.

2. Blacken, insert, turn, and remove the key. If a mark is present below the point you filed, the cut is too shallow. Carefully file it deeper. When the cut is deep enough, use emery paper to clean off the fine burrs left by the file.

3. Reinsert the key and turn it again. It should pass the wards. If it tends to stick slightly, quickly and lightly pass over the cut with the warding file to alleviate the problem. *Note:* The cut should be square on all sides. Think of each ward cut as a miniature square, perfect and even on all sides.

4. Smoke the key, insert it, and turn it. The edges of the cuts should be shiny. The brighter the spot, the greater the pressure at that point. As the individual cuts are filed deeper, the bright spots grow dimmer. When this happens, you are close to the point where you should stop filing.

You must be more cautious each time you insert the key to test the depth of the cut. You are working "in the blind"; a small overcut means the depth is permanently wrong on the blank and you have to start over. This is the reason for making a few light but firm strokes as you near the completion of each cut. Take your time and make one perfect key instead of rushing the job and making many incorrectly cut keys.

The cuts on the key should be as deep and as wide as necessary but no more. Overly large cuts interfere with the action of the lock and may force you to cut another blank.

Once the key is complete, it must be *dressed out*. Use emery paper to remove burrs before they break off and fall into the lock, then polish the entire key with emery paper. For appearance, you might also give the key a light buffing on your wheel. It does you no good to give the customer a dirty and smudged key. Show pride in your work!

Pass keys

Skeleton-type pass keys (master keys) are sold in variety stores. These keys will fit many old locks and more than a few new ones. As such, they are convenient. However, no reputable locksmith stocks them. Why? The ethics of the profession forbid it. You certainly do not want to supply someone with a key that could open his neighbor's lock. Nor should you duplicate a pass key without authorization from the owner. Be leery of a customer who wants a key duplicated but with additional cuts. Locksmiths have lost their licenses for less. Don't let it happen to you.

Lever Tumbler Locks

Lever tumbler locks (or lever locks) have many uses in light security roles. Available in a variety of sizes and shapes, these locks are found on desks, mailboxes, lockers, bank deposit boxes, and other devices.

Figure 5.1 illustrates a popular example of a lever tumbler lock. The circular "window" on the back of the case aids the locksmith by revealing the heights of the levers so you aren't forced to dismantle the lock. Thanks to the window, it is relatively easy to make a key "in the blind."

WINDOW

Figure 5.1 Notice the "window" on the cover of this lever tumbler lock.

Parts

A lever tumbler lock has six basic parts:

- Cover boss
- Cover
- Trunnion
- Lever tumblers (top, middle, and bottom)

- Bolt (bolt stop, notch, and post)
- Base

These parts are shown in Figure 5.2.

Figure 5.2 A lever tumbler lock in exploded view. This model has three tumblers; others might have a dozen or more.

Figure 5.3 Lever tumbler nomenclature. The key bears against the saddle.

Operation

This lock requires a standard flat key. When the key is turned, the various key cuts raise corresponding lever tumblers to the correct height. As the levers are raised, the gates (Figure 5.3) of the levers align and release the bolt. The bolt stop (or post)

Figure 5.4 Some lever tumblers have an open gating.

must pass through the gating from the rear to the front *trap* or vice versa, either unlocking or locking the lock.

Lever tumblers

The number of lever tumblers varies. Most locks have no more than five, although deposit box locks have more. The lever tumbler consists of six parts:

- Saddle (or belly)
- Pivot hole
- Spring
- Gating slot
- Front trap
- Rear trap

Over the years, manufacturers have developed a variety of lever types (Figure 5.4). The operating principle is the same for all of them.

Each *lever* is a flat plate that is held in place by a pivot pin and a flat spring. Each lever has a gate cut into it. The gates are located at various heights with the saddle either aligned for all levers or staggered. The latter approach is antiquated. When the levers are raised to the proper position, the gates are open and the bolt post can shift from one trap to another, thus locking or unlocking the lock.

Because the bolt post meets no resistance at the gating, the lock works properly. On some designs the edge of the lever has serrated notches. The bolt post has corresponding notches. The notches on the lever and the bolt jam together if an improperly cut key is used. This effectively stops the bolt from passing through the gating and keeps the lock secure. Only a perfectly and properly cut key will open this type of lock. This feature adds immensely to the security of the lock.

The width of the gate is a critical factor in the operation of the lock. Some gates are just wide enough for the pin to pass through. A duplicate key, even slightly off on a single cut, will not work on the lock.

The saddle of the lever is also important. Recall from Chapter 1 that staggered saddles make it possible to cut a key by observation. In the case of modern lever tumblers, the gate traps have different heights, which leaves the saddles in perfect alignment.

There are two methods for making gating changes. The most common is to substitute a lever tumbler with a different gate dimension. Locksmith supply houses stock a variety of levers, so all you are required to do is change the original for one with a higher or lower gate. The second method is to alter the tumbler by filing the saddle. This approach is used with levers whose movement is restricted at the gate. Unless the tumbler gating varies greatly, the curvature of the saddle must vary with the shape of the key.

When working on the lever tumbler lock, as on other types, it is best to have two available: one for disassembly to determine internal working parts and one for actual problem-solving.

The typical lever tumbler lock contains two, three, five, or six levers. Bank deposit-box locks can have as many as 14 levers. Lever tumbler locks can be keyed individually, alike (two or more with the same key), or masterkeyed, depending on the wishes of the buyer.

Repairs

General-purpose lever tumbler locks come in three styles. In order of popularity, these are the following:

- Solid case, usually spot-welded or riveted
- Pressed form with the back and sides of the case one piece—small tabs from the sides bend to hold the cover in place
- Cover plate secured by a screw

Riveted or spot-welded locks should be discarded when they fail. It is cheaper to purchase a new lock; the time required to drill out the rivets or chisel through the spot welds costs the customer more than the lock is worth.

Locks secured by tabs can be disassembled and reassembled quite easily. To disassemble, insert a thin tool or small screwdriver under the flanges and pry upward. Remove the cover. Look for a small object jammed in the keyway; if you find something, remove it. At times like this you may wonder if being a locksmith is really worth it, but a locksmith must have patience with the small jobs as well as the big ones.

Broken springs are another common problem with lever tumbler locks. If it is not the top lever, carefully remove each lever in turn, placing them in a logical order so you can reassemble them in the correct order. Remove the lever with the broken spring. Select a piece of spring steel from your inventory. Cut and bend it to shape. If you have purchased an assortment of ready-made springs, select the proper one then replace the broken spring with the new one. Replace the levers and reassemble the lock.

Varieties of Lever Tumbler Locks

Lever tumbler locks are used in a variety of light security applications. Some of them include safe deposit boxes, suitcases, lockers, and cabinets.

Safe deposit box locks

Lever tumbler safe deposit box locks normally have at least 6 and as many as 12 or even 14 levers. These locks require two keys. One key is assigned to the individual who rents the box; the second (or *guard*) key is held by the bank. Both keys are needed to open the lock. If one key could be turned to the open position by itself, it would mean that the key was faulty, the lock mechanism was in some way evaded, or the bolt post was bent or broken.

Disassembly is simple. Remove the screws and lift off the plate. Notice that the lock is constructed differently than previously discussed lever tumbler locks. There are two sets of lever tumblers; two bolt pins must pass through the lever gates at the same time. Notice the unique lever shape (Figure 5.5). The levers have a V-shaped ridge that matches a similar V cut in the bolt post. Another key, even one that is 0.001 inch off on any cut, will mesh the notches in the post and lever. This is another built-in security feature of these locks.

Many safe deposit locks have a compression spring bearing against the upper lever that forces the lever stack down, allowing no play between them. This spring is an integral part of the lock's security mechanism. Without the spring, the levers would be able to move a fraction of an inch when a key or pick "irritates" them. This movement is enough to provide clues for the lock pick artist.

Broken springs are the main difficulty. The levers may also be at fault. The saddles may wear enough to affect the gate. In cases like this, replace the lever with a new one. Do not file the gate cut wider to compensate. Sometimes a pivot post will work loose. Repair it by rapping the post with a light hammer or by brazing. Never use solder. You can straighten slightly bent parts, but it is best to replace the entire damaged part.

Safe deposit locks are serviced by locksmiths who enjoy high standings within the community. Unless an emergency arises, a beginning locksmith or a newcomer to the area would not get the job. This reluctance reflects the conservatism of bankers. It does not reflect on the skills of students and those working locksmiths who have had only a few months' experience. Many of these men and women could repair a safe deposit box lock without difficulty.

Figure 5.5 Notched tumblers are a security feature in some locks.

Suitcase locks

Suitcase locks are essentially simple but have been built to accept a staggering variety of keys. Ninety-nine locks out of a hundred are warded with a primitive bolt mechanism to keep the case closed. Only a few, such as the Yale luggage lock, use a lever-type mechanism. These locks offer better security than warded locks.

The lock size, the depth the key is inserted, and the number of cuts are important clues to the lock type. A key for a lever tumbler lock will go ¼ inch or deeper into the lock before turning. Warded locks are shallower.

Opening suitcase locks is relatively simple. Many times one key will open several different suitcase and luggage locks. Many of these locks on the market are not designed for security. By cutting down almost any suitcase key, it is possible to make a skeleton key for emergencies.

Locker and cabinet locks

The Lock Corporation of America. The Lock Corporation of America has come up with a new locking spring design accompanied by a free-turning keyway that combine to make the locks extremely jamproof and pick resistant for a flat-key lock. Figure 5.6 shows such a lock.

These locks are made to last (they carry an unconditional guarantee). They are constructed of a one-piece, heavy-duty steel case that will stand up to years of use, even under the most corrosive conditions. The hard-wrought steel keys and the lock case are electroplated and dichromated to ensure numerous years of trouble-free service.

Another advantage to these locks is that they were designed to fit into current master key systems. The lock will fit left- or right-hand steel locker or cabinet doors that have the standard piercings.

The keys can be removed from either the deadbolt or springbolt locks when the locker door is opened or closed. The deadbolt locks are available in models with the key removable in the locked or unlocked mode or with the key removable in the locked position only. Also, with these locks, you have a self-locking springbolt with the key removable in the locked position only.

Figure 5.6 A lever tumbler deadbolt lock for lockers. (*Lock Corporation*)

Figure 5.7 Notice how a lever tumbler lock can be installed on a locker. (*Lock Corporation*)

Series 4000 and 5000. The series 4000 and 5000 flat key built-in lever tumbler (Figure 5.7) specifications are as follows:

- Mounts in a standard three-hole piercing on steel lockers, cabinets, and other shop equipment
- Solid-locking bolt moves horizontally with a ⁵⁄₁₆-inch throw
- Masterkeyed, keyed different, keyed alike, or group keyed
- Key removable when locked, or when locked or unlocked (must be specified when ordered)
- ⅝-inch-gauge, heavy-duty, one-piece steel case
- Free-turning keyway
- Phosphor bronze locking springs
- Five hard-wrought steel levers

The dimensions for the lock and the standard locker door punchings are shown in Figure 5.8.

The model 5001 MK locker lock is a built-in, flat-key lock with a deadbolt locking mechanism; the key is removable in either the locked or unlocked position. A free-turning plug (with a minimum of five lever tumblers) provides a minimum of 1400 different key changes. The locks (within a key change range specified) can be masterkeyed if desired. The lock will not be operated by keys of another lock within the specified key range, except a master key. The casing is of wrought steel and is 1⅝ inches × 1⅝ inches with top and bottom attaching ears to fit the standard steel locker piercings. All parts are of phosphor bronze, zinc, or rustproofed steel. The backset is 1 inch or less, and the solid locking bolt is ¾ inch × ¼ inch with a ⁵⁄₁₆-inch throw.

Figure 5.8 Specifications for the LCA series 4000 lever tumbler lock. (*Lock Corporation*)

Figure 5.9 Internal parts for the 4000 and 5000 series lever tumbler lock. (*Lock Corporation*)

Specifications are the same for the 5000 MK except for a key removable in the locked position only; the 4000 MK differs only in having a solid springbolt locking mechanism. The internal parts' relationship and positioning are illustrated in Figure 5.9.

Torsion lever locks. The LCA torsion lever flat-key lock contains a patented, multi-movable lever, flat-key operated mechanism that has the latest improvements to the world's oldest and still most popular means of security—the warded lock.

Seven levers (including one master lever, if required) engage and restrain the locking bolt. The novel uni-spring construction is arranged in two banks of seven staggered, interlocking arms constantly exerting a force on all the levers. When you insert and turn the precision-cut key in the rotatable key guide, the latched levers move from securing the deadbolt, enabling the lock to open.

Note: the model 4000 (basically the same) has a bolt split into two parts, including a separate spring that allows the bolt to retract without using a key to close or slam shut.

The grooved key, built-in torsion tumbler cylinder lock (Figure 5.10) has the following specifications:

- Standard three-hole piercings for mounting
- Beveled locking bolt, moving horizontally
- Locking bolt is ¾ inch × ⅜ inch with a ⁵⁄₁₆-inch throw
- Over one million computerized key changes
- Eight nickel silver torsion tumblers (to provide maximum security)
- Regular, submaster, masterkeyed, keyed different, keyed alike or group keyed
- The master key design prevents the use of a regular key as a master key
- The keys are embossed, nickel plated, and grooved
- Cylinder plug of solid brass
- ⅝-inch-gauge heavy-duty one-piece steel case construction

The torsion tumbler is based on a new and different design concept. Each of the eight nickel silver torsion tumblers is manufactured with its own integral spring; it is

Figure 5.10 Built-in torsion tumbler cylinder lock. (*Lock Corporation*)

made strong and durable by an exclusive forging process. This tumbler-spring design provides a much greater protection against unfavorable climatic conditions such as temperature and humidity changes because the torsion spring is heavier and sturdier than the finely coiled springs required by either pin or disc tumbler cylinders.

Variation in the eight tumblers (arranged four to each side of the cylinder) and in the double-bitted key provide more than 1 million key change possibilities. The unique open construction of a one-piece tumbler and spring eliminates clogging due to accumulated dirt and small particles, and offers long reliability, high-performance, smooth operation, and improved pick resistance.

Better master key control and protection are provided by an entirely different master key configuration. This special key design prevents the fashioning of a master key from regular keys or blanks, thus reducing the possibility of unauthorized entry. Practically unlimited master key changes are available without diminishing the security of the cylinder.

Figure 5.11 provides details of the maximum security eight torsion tumbler cylinder in an exploded view. Figure 5.12 shows the critical dimensions of the cylinder.

Series 6000 and 7000.   The 6000 and 7000 series have a patented snap-on face plate for installation (Figure 5.13). Technical drawings for the snap-on installation are in

Figure 5.11 Exploded view of the maximum-security, 8-torsion tumbler cylinder. (*Lock Corporation*)

Figure 5.12 Critical cylinder dimensions of the 8-torsion tumbler cylinder lock. (*Lock Corporation*)

Figure 5.13 Series 6000 torsion tumbler lock with inset illustrating the snap-on face plate. (*Lock Corporation*)

Figure 5.14, along with the locker door hole punchings. Note that the series 6000 has a springbolt, while the 7000 operates with a deadbolt.

The 7001 locker lock has a built-in high-security type of ruggedness with a deadbolt locking mechanism and the cover plate; it is masterkeyed with the key removable in either the locked or unlocked position. Over 100,000 possible different key changes are available, The bolt is ¾ inch × ⅜ inch with a ⁵⁄₁₆-inch throw.

The LCA drawer and cabinet lock (Figure 5.15) is constructed of high-pressure solid Zamak unicast casing and base and has a ⅞-inch diameter cylinder case for use in up to 1⅜-inch-thick material.

Push locks. The LCA push lock for sliding door and showcases uses the same patented torsion tumblers and solid brass cylinder plug and it has a stainless steel cap (Figure 5.16). Three hundred key changes are standard, but unlimited changes are available.

The various series 3300 and 3400 locks can have a push bolt, automatically locked

"SNAP-ON" installation

$3/8 \times 50°$

$1/4$"

On

$5/16$" Throw

$7/32$"

For
$7/32$"
hole

$1^{23}/32$"

$55/64$"

$45/64$"

$3/4$"

$3/8$"

$2^3/16$"

$1^3/32$"

Standard locker
door punchings

$7/32$"

$21/32$"

$2\ ^3/16$"

$1\ ^3/32$"

Figure 5.14 Specifications for
series 6000/7000 "snap-on"
installation lock. (*Lock
Corporation*)

by pushing the cylinder inward or releasing the cylinder and catch by turning the key 90°. The bolt is ⅜-inch diameter with a ¹⁵⁄₃₂-inch throw and a ⅞-inch cylinder case for use in up to 1⅛-inch-thick material mountings.

Figure 5.15 An 8-torsion tumbler cylinder drawer lock. (*Lock Corporation*)

Figure 5.16 Sliding door and showcase locks. (*Lock Corporation*)

The push-turn bayonet type lock in this LCA series has a different strike plate (Figure 5.17). It is locked by turning the key 90°, pushing the cylinder in and rotating the key again to lock the T type plunger behind the strike plate, thus preventing the doors from being pried apart. The cylinder locks in either the in or out positions for further protection. With the 3350 model lock, a ⁷⁄₁₆-inch-diameter, case-hardened, steel-covered bolt is available. Figure 5.18 shows the 3300 model technical dimensions, while Figure 5.19 provides the same information on the 3400 model.

Figure 5.17 Grooved key push lock for sliding doors and display/showcases. (*Lock Corporation*)

Figure 5.18 Technical dimensions for the sliding door pushbutton lock. (*Lock Corporation*)

Figure 5.19 Technical dimensions for the push and turn showcase lock. (*Lock Corporation*)

For the locksmith concerned with businesses, these locks are a stock in trade, constantly proving themselves over and over.

Lever Tumbler Lock Keys

Unlike the warded or cylinder key, the lever tumbler lock key used by the average person almost never has a keyway groove running along its side; it is a flat key. A number of different flat keys exist. Figures 5.20 and 5.21 show some of them. Figure 5.22 identifies the various parts of a typically cut lever tumbler lock key.

Cutting keys

The lever tumbler lock key can be cut by hand or machine. To make the key, you must have the proper key blank. The three critical dimensions of the lever tumbler lock key are the thickness, length, and height. If the key blank is slightly higher or wider than the original, the blank should be filed down to the proper size. If it is thicker, select another blank. Filing down the thickness of the blank weakens it structurally.

Lay the keys (original and blank) side by side, then run your finger lightly across them. If the blank is thicker than the original, you will feel your finger catch as it passes from one to the other. When you insert the key into the keyway, it should not fit tightly or bind.

The first cut is called the *throat cut*, which enables the key to turn within the keyway. To make a throat cut, insert the blank in the keyway and scribe each side of the

Figure 5.20 Typical flat key blanks used to make keys for lever tumbler locks.

Figure 5.21 Look-alike key blanks for lever tumbler locks.

Figure 5.22 Lever tumbler lock key nomenclature.

blank where it comes in contact with the cover boss. Determine the point where the trunnion or pin of the lock turns. Draw a vertical line there to indicate the depth of the throat cut. Remove the key from the lock, place it in a vise, and use a 4-inch warding file to cut alongside the vertical line to the proper depth. Cut on the side of the line toward the tip of the key.

There is a small round window in the back of most lever tumbler lock cases. The window is positioned so you can see where the bolt pin meets the lever gates. By observing the lever action through the window, you can get a general idea of the proper cuts to make on a key blank.

After the throat cut is filed, make the other cuts.

1. Smoke the blank, insert it in the keyway, and turn. Remove it. The lever locations are marked on the blank.

2. File the marks slightly, starting with the marks closest to the throat cut.

3. Insert the key and turn it. Notice (through the window) the height to which each lever comes up and the position of the pin in relation to the lever gates. The distance from each gate to the pin indicates the cut depth for each lever. File each key cut a little at a time, periodically inserting and turning the key to check the gate/pin relationship. Be sure to file thin cuts. Continue until you can insert the key and have the gate and pin line up exactly.

If the key in the keyway binds, observe the levers. One or more may not be correctly aligned. If the pin is too high, you have cut too deeply; if the pin is too low, you have not cut deeply enough. It may be necessary to resmoke the blank and reinsert it. The point where the key binds hardest will be indicated by the shiniest spot on the key. Just a touch with the file will usually alleviate the problem. Ensure that each filed cut is directly under its own lever.

At this point, you should correct any variations between the original key and the blank. The dimensions must be identical: the key height, thickness, and length.

You need a vise for holding the two keys, a small C-clamp, a warding file, a candle, and pliers. Follow this procedure:

1. Hold the original key over a candle flame and smoke it thoroughly.

2. Allow a few minutes for the key to cool; clamp the original and the blank at the bows. Most locksmiths use a C-clamp for this initial alignment.

3. Once aligned, place the key and blank in a vise. If you wish, you can leave the C-clamp in position.

4. Use the warding file to make the tip cut first. File in even and steady strokes, bearing down in the forward or cutting stroke. Keep a careful eye on the original. Stop when the file just disturbs the blackening.

5. Once the tip cut is completed, move to the next cut.

6. Remove the keys from the vise and inspect the cuts. Each should be rectangular and flat.

7. Use emery paper to lightly sand away the burrs on the edges of the cuts. Wipe off the candle black from the original key.

8. Test the duplicate in the lock. Should it stick, blacken the duplicate and try it again in the lock. Breaks in the blackening will show where the key is binding. A light stroke with the file should correct this.

Reading lever tumbler locks

After you learn how to read lever tumbler locks, you'll be able to make working keys without disassembling the locks. As you know, a lever tumbler lock keyway is narrow, and the view of the tumblers is further restricted by the trunnion. While this is a handicap, it can, in part, be overcome with the help of an appropriate reading tool.

The positions of the lever tumbler saddles are one clue to reading the lock. The saddle of the lock can tell you quite a bit. The wider the saddle, the deeper the cut; the narrower the saddle, the shallower the cut. Using the reading tool, you will feel the different saddle widths to determine the cuts and develop some idea of the cuts and of the key shape.

To do this properly, you must have an appreciation of the internal workings of the lock. Depth key sets can be useful when you are ready to cut the keys. Lever cuts are usually in 0.015–0.025-inch increments.

Practice is required. Begin with a lock that has a window. If possible, obtain extra levers so you can change the keying of the lock at will.

Raise each lever with the reading tool and determine the proper position for each one. Remember the general rule: a wide saddle requires more movement than a narrow saddle. Once you are satisfied that you have read the lock, disassemble it and examine the lever tumblers. You may have misread the tumblers but do not be disheartened. The only way to achieve competence in this skill is to practice.

Other tumblers are designed with uniform saddle widths. Keying is determined by the positions of the gates in the tumblers. These locks can be quite difficult to read.

Working with the reading tool, raise one of the levers as high as it will go. While this movement is not a direct indication of the depth of the key cut, it is important. The amount of upward movement establishes the minimum key cut; a shallower cut will jam the levers against the top of the lock case. In addition, the individual gates and traps are usually in some rough alignment. You will usually find that two are in almost perfect alignment.

A cut for a lever that has the post on the upper half of the trap will be shallow; a cut for a lever that has its post on the lower half will be deep. A lever that has its gate in the intermediate position will require a key cut between these extremes.

Once you have established the general topography of the cuts, refer to your set of depth keys for the exact dimensions. Established locksmiths have reference manuals that may simplify the work.

Chapter

6

Disc Tumbler Locks

The disc tumbler (or cam) lock gets its name from the shape of the tumblers. These are about as secure as lever locks and as such are superior to warded and other simple locks. However, disc tumbler locks are far less secure than pin tumbler locks.

Disc tumbler locks are found in automobiles, desks, and a variety of coin-operated machines. Some padlocks are built on this principle. Because the cost of manufacture is very low, replacement is cheaper than repair.

While similar to the pin tumbler lock in outside appearance and in the broad principle of operation, the internal design is unique.

The disc tumblers are steel stampings arranged in slots in the cylinder core. Figure 6.1 shows a typical disc tumbler. The rectangular *hold* (or *cutout*) in the center of the disc matches a notch on the key bit. The protrusion on the side, known as the *hook*, locates the spring. The disc stack is arranged with alternating hooks, one on the right of the cylinder and one on the left. The complete lock is illustrated in Figure 6.2.

Figure 6.1 Disc tumbler lock in cross-section. The position of the cutout determines the depth of the key cut.

First tumbler   Second tumbler

No key inserted

Wafers protrude past plug shear lines

Double-sided disk tumbler key

Figure 6.2 Parts of a common disk tumbler desk lock and a typical key for such a lock. (*Rathjen,* Locksmithing: From Apprentice to Master, *1995 McGraw-Hill, Inc., reproduced with permission*)

Operation

The disc lock employs a rotating core, as does the more familiar pin tumbler design. The disc tumbler core is cast so the tumblers protrude through the core and into slots on the inner diameter of the cylinder. So long as the tumblers are in place, the core is locked to the cylinder.

The key has cuts that align with the cutouts in each tumbler. The key should raise the tumblers high enough to clear the lower cylinder slot but not so high as to enter the upper cylinder slot. In other words, the correct key will arrange the tumblers along the upper and lower shear lines (Figure 6.2). The plug is free to rotate and, in the process, throw the bolt.

The key resembles a cylinder pin-tumbler key except that it is generally smaller and always has five cuts. A cylinder pin-tumbler key might have six or seven cuts.

Disassembly

Good quality disc locks have a small hole on the face of the plug, usually just to the right of the keyhole. To disassemble the lock, insert a length of piano wire into the hole and press the retainer clip. Turn the plug slightly to release it. The key gives enough purchase to withdraw the plug. If a key is not available, you can extract the plug with the help of a second length of piano wire inserted into the keyhole. Bend the end of the wire into a small hook. Other locks attach the plug to the cylinder with the same screw that secures the bolt-actuating cam. Others (fortunately, a minority) have the plug and cylinder brazed together. File off the brass.

Keying

Manufacturers have agreed on five possible positions for the cutouts relative to the tumblers. Keying is a matter of arranging the tumblers in a sequence that matches the key cuts. Once the sequence has been identified, install the tumbler springs over their respective hooks and mount the tumblers in the plug. The tumblers are spring-loaded and, until the plug is installed in the cylinder, are free to pop out. Lightly stake them in place with a punch or the corner of a small screwdriver blade. One or two pips are enough because you will break the tumblers free once the assembly is inside the cylinder. Inserting the key is enough to release the tumblers.

Security

These locks have no more than five tumblers, and each tumbler cutout has five possible positions. These variations allow 3125 or (5^5) key changes. In practice, the manufacturer will discard some combinations as inappropriate and may further simplify matters by limiting the key changes to 500 or less. Obviously, disc tumbler locks are not high-security devices.

Cam Locks

Cam locks are used for a variety of general and special purposes (Figure 6.3). More than likely, you will see approximately 90 percent of these units in offices. From the *face* (or front) these locks look pretty much the same after they have been installed. Inside, however, they may differ in various regards. The illustrations on the next several pages are exploded views of different models and the parts associated with each. These models are all five-disc tumbler locks, each with a possibility of 200 different key combinations. Because of the key variations possible, these locks are often masterkeyed prior to purchase by the customer. Only by disassembling the lock or viewing the tumblers through the keyway will you know for sure whether or not the lock is masterkeyed (Figures 6.4 through 6.9).

Two other core-type cam locks are shown in Figures 6.10 and 6.11. These cam locks with removable cores are of the seven-pin variety that uses the Ace-type circular key.

Figure 6.3 Disc tumbler cam lock. (*Dominion Lock Co.*)

Parts
Desk lock

1. Nut
2. Bolt
3. Shell
4. Retainer clip
5. Spring
6. Plug
7. Springs
8. Disc tumblers
9. Key

Figure 6.4 Exploded views of a disc tumbler cam lock for desks. (*Dominion Lock Co.*)

Parts
Cam locks

1. Nut
2. Lock washer
3. Cam
4. Stop cam
5. Nut
6. Shell
7. Plug
8. Springs
9. Tumblers
10. Key

Figure 6.5 Exploded view of cam lock. (*Dominion Lock Co.*)

A variety of cams is used with these locks. Figure 6.12 is a chart showing the various types of cams available. The length and offset specifications are also included. Figure 6.13 shows the hook, bent, double-ended, and other miscellaneous cams that you may come across in the course of your work.

The standard thickness of a cam is ³⁄₃₂ inch (2.667 mm). They are made of steel; many are cadmium-plated for durability and longevity.

Reading Disc Tumbler Locks

It's not unusual for a locksmith to be asked to make a key for a lock whose original key was lost or misplaced. If the lock does not have a code number on its face or if the owner neglected to write down the number on the key, the locksmith has three choices. He can pick the lock, impression a key, or "read" the lock.

Like impressioning, reading a lock is a skill that must be developed through patient practice. You cannot expect to master this skill quickly, nor can you expect to remain proficient in it without constant practice. At first, practice daily, then weekly or twice-weekly to maintain your skill.

When called on to fit a key, either by impressioning or reading of the lock, the locksmith invariably looks into the lock keyway. A quick glance determines whether it is a lever, disc, or pin tumbler lock.

Disc locks

The view down the keyway of a disc tumbler lock will show a row of discs with their centers cut away and staggered. The cutaways are at different heights; the discs themselves are the same diameter. The cutaway looks like a small staircase with a

Parts
Drawer lock

1. Mounting screw
2. Bolt
3. Shell
4. Retainer clip
5. Spring
6. Plug
7. Springs
8. Disc tumblers
9. Key

Figure 6.6 Drawer lock, exploded view. (*Dominion Lock Co.*)

Parts
Drawer lock
with removable backplate

1. Mounting screws
2. Removable backplate
3. Bolt
4. Shell
5. Retainer clip
6. Spring
7. Plug
8. Springs
9. Disc tumblers
10. Key

Figure 6.7 Drawer lock, exploded view. (*Dominion Lock Co.*)

Parts
Sliding door lock

1. Mounting screw
2. Case
3. Screw
4. Spring
5. Shell
6. Spring
7. Retainer clip
8. Spring
9. Cam
10. Plug
11. Springs
12. Disc tumblers
13. Key

Figure 6.8 Sliding door lock, exploded view. (*Dominion Lock Co.*)

Parts

Sentry series cam lock

1. Screw
2. Lock washer
3. Cam
4. Stop cam
5. Nut
6. Shell
7. Disc tumblers
8. Springs
9. Plug
10. Key

Figure 6.9 Sentry series cam lock, exploded view. (*Dominion Lock Co.*)

Parts

Cam locks

1. Nut
2. Lock washer
3. Cam
4. Stop cam
5. Cylinder
6. Springs
7. Top pins
8. Bottom pins
9. Spindle assembly
10. Nut
11. Shell pin
12. Shell
13. Key

Figure 6.10 Tubular key cam lock, exploded view. (*Dominion Lock Co.*)

Parts
Removable core lock

1. Cam
2. Sleeve
3. Stop cam
4. Cylinder
5. Springs
6. Top pins
7. Bottom pins
8. Spindle assembly
9. Shell pin
10. Shell
11. Key

Figure 6.11   Removable core lock, exploded view. (*Dominion Lock Co.*)

surrounding wall (formed by the vertical edges) around the steps. Since each disc is the same size, only the height of the "steps" varies, and this variation is predetermined. A No. 1 disc has its cutaway toward the top of the tumbler; a No. 5 has its cutaway situated low on the tumbler (Figure 6.14).

The skill of reading a disc lock is learned here. Study the discs through the keyway. Notice the relationship of the discs to each other and to the keyhole. Through constant study and practice, which includes mixing the various discs, you will be able to determine which disc is which (Figure 6.15).

Lift each disc and compare it to the disc in front of or behind it. To do this, you will need a *reading tool*. This is nothing more than a stiff length of wire about 3 inches long, mounted in a small dowel handle. Think of the tool as a long hairpin attached to a short piece of wood for convenience in holding. The wire should be bent slightly so you can see the tumblers.

Insert the tool into the lock, holding it so you can see the interior and observe the discs. By shifting the tool to raise and lower each disc, you can see the relationship of each disc cutaway to the next disc and to the keyway. Using your knowledge about the cutaway relationships, you will be able to decode the tumblers and cut a key for the lock.

You might ask yourself, "How do I know where to cut the key—and how deep?" Recall that when impressioning a key, you blackened the key and determined the cutting point by the pressure of the levers on the key as you tried to turn it. The same

Figure 6.12 Cam lengths and offsets. (*Dominion Lock Co.*)

LENGTH			
	A	2 3/8"	60.325 mm
	B	2 1/4"	57.150 mm
	C	2 1/8"	53.975 mm
	D	2"	50.800 mm
	E	1 7/8"	47.625 mm
	F	1 3/4"	44.450 mm
	G	1 5/8"	41.275 mm
	H	1 1/2"	38.100 mm
	I	1 3/8"	34.925 mm
	J	1 1/4"	31.750 mm
	K	1 1/8"	28.575 mm
	L	1"	25.400 mm
	M	7/8"	22.225 mm
	N	3/4"	19.050 mm

OFFSET			
	O	1/2"	12.700 mm
	P	3/8"	9.525 mm
OUTSIDE	Q	1/4"	6.350 mm
	R	3/16"	4.775 mm
	S	1/8"	3.175 mm
	T	1/16"	1.600 mm
	U	1/16"	1.600 mm
	V	1/8"	3.175 mm
INSIDE	W	3/16"	4.775 mm
	X	1/4"	6.350 mm
	Y	3/8"	9.525 mm
	Z	1/2"	12.700 mm

technique applies here. Insert a blackened key and give it a slight turn. This brings the key in contact with the sides of each tumbler cutout. The cutout leaves a mark on the blank, indicating the portion of each cut.

Determining the depth of the cuts requires experience. You have already learned that cutting a key is slow, patient work. As you get closer to the proper depth, move the file with less pressure than before. The same care and precision is needed here. A disc with a No. 1 cut hole requires a deep key cut as compared to a No. 4 or 5 disc, which require a very shallow cut.

For reference aids, use various extra keys that you have collected. Since the depths of the cuts are standard in the industry, you can look at a key with a 13354

Figure 6.13 Hook, offset, double-ended, and miscellaneous-type cams. (*Dominion Lock Co.*)

Figure 6.14 Disc tumblers have five variations.

Figure 6.15 The position of the tumblers gives a general idea of the key profile.

cut and, by observing the differences in the depths of each cut, know exactly how deep a No. 2 cut should be. Obtain a depth key set from a locksmith supply house. This set has a different key for each depth, with the same cut in all five tumbler positions. Thus, a No. 2 key has five No. 2 cuts; a No. 3 key has five No. 3 cuts; and so on through the series. With the help of these keys and your trained eyesight, you can place a blank in a vise and cut the key by hand.

Without a depth key set, you can use a variety of disc keys as guides. Select a key with the proper cut and align your blank to it. You can also make your own set of depth keys. Making the set teaches you how deep to make each cut at any given position on the blank.

Double-Bitted Disc Tumbler Locks

The Junkunc Brothers American double-bitted disc tumbler lock is usually found in padlocks and office and utility locks (Figure 6.16).

Operation

When the double-bitted key is inserted, it passes through the center of the tumblers (as in a disc lock) and aligns them to the shear line, allowing the plug to rotate. But, the key and the tumbler arrangement is different from that of the regular disc lock.

Figure 6.16 Double-bitted padlock and key. (*American Lock Company*)

The key cuts are wavy in appearance; thus the tumblers have to align in a wavelike configuration for the lock to open. Further, there are no definitive tumbler cuts on the key. This is because the key holds ten or more tumblers compressed together and held in a locked position by means of a Z-shaped wire within the tumblers.

All the tumblers are *uncoded*, meaning they are all of a standard cut. For them to turn within the cylinder, the tumblers have to be cut down. A special keying tool is used for this purpose.

Cutting down the tumblers

1. Insert the tumblers into the plug (with the tumbler spring in place), then insert the precut key into the plug.

2. Mount the plug firmly in your vise.

3. Attach the keying tool to a ¼-inch drill. Drill the back of the plug. Since the inside diameter of the drill is the same as the outside diameter of the plug, the individual tumblers will cut down to what will be the shear line.

4. Trim the tumblers with a light wire brush to take off any burrs.

5. Insert the plug into the cylinder and test it. Attach the retainer screw and withdraw the key.

6. Since they are uncoded, the tumblers can be used within any plug. The tumbler spring, because of the shape, holds the various tumblers in the locked position. Only with the insertion of a key, which forces the tumblers into another position, can the lock be opened.

Keys and keyways

The double-bitted lock takes four basic key sections (Figure 6.17). These sections, of course, match the shapes of the keyways. Keyway one is referred to as a K4 and the center point is at the center of the tumbler. Keyway two is referred to as a K4L; the center point is just left of center. Keyway three is called a K4R; the center point is right of center. Keyway four, called a K4W, is shaped like a W. The keyway shape does not reflect the tumbler types that are within any given plug.

Figure 6.17 Four basic double-bitted key sections. (*Taylor Lock Company*)

7

Pin Tumbler Locks

There is a wide variety of pin tumbler locks, including deadbolt, key-in-knob, lever, padlock, automobile ignition lock, etc. The common denominator among pin tumbler locks is that they all have a pin tumbler cylinder or housing.

When a pin tumbler lock is installed on a door you can usually see only the lock's plug or the face of its cylinder. Figures 7.1, 7.2, and 7.3 show examples of common pin tumbler locks. You can quickly identify a pin tumbler lock by looking into its keyway. You'll see one or more pin tumblers hanging down.

Construction

Although pin tumbler cylinders are simple mechanisms, some of the most secure mechanical locks rely on such a cylinder. Most pin tumbler cylinders are self-contained mechanisms. They come in a variety of shapes to fit locks of various designs. Figures 7.4, 7.5, and 7.6 show some differences among pin tumbler cylinders.

The basic parts of a pin tumbler cylinder include: a cylinder case (or shell); plug (or core); keyway; upper pin chambers; lower pin chambers; springs; drivers (or top

Figure 7.1 Pin tumbler cylinder key-in-knob lock. (*Schlage Lock Company*)

Figure 7.2 Pin tumbler cylinder deadbolt lock. (*Schlage Lock Company*)

Figure 7.3 Pin tumbler cylinder mortise lock. (*Omnia Industries, Inc.*)

Figure 7.4 Pin tumbler cylinder for rim locks. (*Ilco Unican Corp.*)

Figure 7.5 Pin tumbler cylinder for key-in-knob locks. (*Ilco*)

Figure 7.6 Pin tumbler cylinder for mortise locks. (*Ilco*)

pins); and bottom pins. It's easy to remember all those parts once you understand their relationships to each other.

The *cylinder case* houses all the other parts of the cylinder. Figure 7.7 shows the parts of a pin tumbler cylinder. The part that rotates when the proper key is inserted is called the *plug*. The *keyway* is the opening in the plug that accepts the key. The drilled holes (usually five or six) across the length of the plug are called *lower pin chambers*; they each hold a *bottom pin*. The corresponding drilled holes in the cylinder case directly above the plug are called *upper pin chambers*; they each hold a *spring* and a *driver*.

Springs, top pins bottom pins

Mortise cylinder shell

Plugs

Cam

Figure 7.7 Cutaway view of face of a pin tumbler cylinder. (*The Locksmith Guild*)

How pin tumbler cylinders work

When a key is not inserted into the cylinder, the downward pressure of the springs "drives" the drivers (top pins) partially down into the plug to prevent the plug from being rotated. Only the lower portions of drivers are pushed into the plug, because the plug holds bottom pins. There isn't enough room in a lower pin chamber to hold the entire length of a driver and a bottom pin.

There is a small amount of space between the plug and the cylinder case. That space is called the *shear line*. Without a shear line, the plug would fit too tightly in the cylinder case to rotate. When a properly cut key is inserted, it causes the top of all the bottom pins and the bottom of all the drivers to meet at the shear line. While the pins are in that position, the plug is free to rotate to the open position (Figures 7.8 and 7.9).

Figure 7.8 Cutaway view of side of a pin tumbler cylinder. (*The Locksmith Guild*)

A   **B**

Figure 7.9 The cylinder shown in (A) employs standard drivers; it has been picked and the plug is free to rotate. The cylinder shown in (B) uses a spool tumbler; it has also been picked, but can be rotated only a fraction of an inch before the spool tumbler jams. (*Allen, Locks & Alarms, 1985 TAB Books, reproduced with permission of McGraw-Hill, Inc.*)

Plug. The plug contains a keyway, usually of the paracentric (off-center) type. Although they usually contain five or six lower pin chambers, some plugs have four or seven. The lower pin chambers are spaced fairly evenly along the upper surface of the plug, and they are aligned as closely as modern production techniques allow. The plug may be machined with a shoulder at its forward surface; this shoulder mates with a recess in the cylinder and provides

- A reference point for regulating the alignment of pin chambers in the case and the plug.
- A safeguard to prevent the plug from being driven through the cylinder, either deliberately or through resistance developed as the key enters the keyway.
- A safeguard to discourage a thief from shimming the pins. Without this shoulder it would be possible to force the pins out of engagement with a strip of spring steel.

The plug is retained at the rear by a cam and screws, a retainer ring, or a driver that locks the plug into the cylinder.

Pins. Top and bottom pins are usually made of brass. They come in a variety of lengths, diameters, sizes, and shapes (Figure 7.10). Although they're small, pin tumblers are very strong.

The shape of the pins helps resist attempts to pick the lock. A standard cylindrical driver easily can be lifted to the shear line with a pick while a tension wrench is used to apply a little turning force to the plug. However, a top pin with a broken profile will tend to hang up before it passes the shear line, which makes a lock more difficult to pick (Figure 7.11). (To learn more about lock picking see Chapter 17.)

Tailpieces and cams. Most pin tumbler cylinders have either a tailpiece or a cam attached to the rear of the plug (Figure 7.12A and Figure 7.12B). The tailpiece is loose to allow some flexibility in the location of the auxiliary lock on the other side of the door. Alignment should be as accurate as possible; under no circumstances should the tailpiece be more than ¼ inch off the axis of the plug.

On the other hand, pin tumbler cylinders for mortise locks do not have such tolerance. The best is driven by a cam on the back of the cylinder (Figure 7.12C). If the lock is used on office equipment, the cam is probably a milled relief on the back of the plug; it might also be a yokelike affair, secured to and turning with the plug. It is important that these locks are aligned with the bolt mechanism.

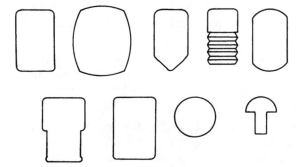

Figure 7.10 Various pin tumblers.

Figure 7.11 High-security top pins "hanging up" below the shear line. (*DOM Security Locks*)

Disassembly

To disassemble a pin tumbler cylinder, you need a small screwdriver, a *following tool* of the correct diameter, and a pin tray. Proceed as follows:

1. Turn the plug about 30 degrees. That is done with the key, by picking the cylinder, or by shimming the cylinder.

2. Remove the cam or tailpiece by removing two screws or a retainer ring. (Although they're not essential, a small pair of *snap ring pliers* can be useful for removing retainer rings.)

3. While holding the cylinder so the pins are in a vertical position, slip the appropriate size plug follower tool into the cylinder directly behind the plug. Use the plug follower tool to slowly push the plug out of the face of the cylinder. Do not create a gap between the plug and the plug follower tool or the springs and drivers will fall out.

4. Hold the plug as it leaves the cylinder; don't let the bottom pins fall out of the plug. Use your index finger to cover all pin chambers in the plug except the one closest to the face of the plug. Empty that chamber and put the pin into the first compartment of your pin tray.

5. Uncover the next pin chamber in the plug and empty the pin into the corresponding compartment of the pin tray. Continue this procedure until all the bottom pins are out of the plug and they are in proper order in the pin tray. Then you can put the plug down for awhile.

6. Hold the cylinder case and slowly pull the plug follower out the back of the cylinder. Stop each time a driver and spring fall from an upper pin chamber. Place them in the tray directly above each corresponding bottom pin.

Figure 7.12 Medeco pin tumbler cylinders. The tailpiece can be serrated (A), solid (B), or replaced by a cam (C). (*Medeco Security Locks*)

Assembly

To assemble a pin tumbler cylinder, you need a plug follower, a small screwdriver, and pin tweezers. A *plug holder* is optional; it holds the plug in an upright position, leaving your hands free while you insert the bottom pins.

This procedure assumes that you have a key but do not know the pin combinations. (In normal reassembly, you already know which pins go into which chambers and you simply insert the springs and pins in the proper chambers.)

1. Place the plug into the plug holder and insert the key into the plug's keyway.

2. Taking one bottom pin at a time, insert it into the lower pin chamber nearest the face of the plug. If the pin stands above or below the shear line, try another pin. When a pin appears to be just at the shear line, turn the plug with the key. If the plug turns 360 degrees in the plug holder, the bottom pin is probably the right one.

3. Repeat the procedure for each lower pin chamber until all the bottom pins just reach the shear line. *Note:* Do not remove the key or turn the plug over.

4. Move to the cylinder case and insert the plug follower through the back of the cylinder until it covers the third upper pin chamber from the face of the cylinder. Turn the cylinder case upside down so the upper pin chambers are facing the floor.

5. Using your tweezers, drop a spring into the second pin chamber from the face of the cylinder. Then, use your tweezers to insert a driver on the spring. By pushing the driver into the pin chamber while applying slight forward pressure on the plug follower, the plug follower will bind the driver into place. That will allow you to release the driver with your tweezers and use your tweezers to push the pin into the upper pin chamber while simultaneously pushing the plug follower over the driver. The plug follower will hold the driver and spring in place so they don't pop out of the cylinder. Repeat the procedure with the first upper pin chamber from the face of the cylinder.

6. At this point, you should have two upper pin chambers filled and covered by the plug follower. Push the pin follower out of the face of the cylinder to the point where your plug follower is covering only the upper pin chambers you've already filled. The remaining upper pin chambers should be exposed.

7. Insert a spring and driver in the exposed upper pin chamber nearest the plug follower. Cover that pin chamber with your plug follower. Repeat this step until all the upper pin chambers are filled.

8. Turn the cylinder case upright so the drivers and springs are in a vertical position.

9. Slowly remove the key from the plug so the bottom pins don't pop out of the plug. Insert the plug into the cylinder by pushing the back of the plug against the plug follower from the face of the cylinder. After the plug has forced the plug follower out of the rear of the cylinder case, rotate the plug around until the upper pin chambers are aligned with the lower pin chambers.

10. The pins should lock the plug into place. *Note:* When inserting the plug, be sure the lower pin chambers are at least 30° out of alignment with the upper pin chambers.

11. Make sure your key works by inserting it into the plug and rotating the plug. Be careful while rotating the plug and removing the key; only the pins are holding the plug in place. With the key in the plug, all the pins should be at the shear line and the plug can easily be pulled out of the cylinder. If that happens, all the drivers and springs will fall out.

12. Attach the cam or tailpiece to the cylinder and retest your key.

Choosing a Pin Tumbler Lockset

Pin tumbler cylinder locksets are the most popular types of locking devices. These locksets are available in a wide variety of types and styles. Some of the general characteristics of such locksets include the following:

Security. The pin tumbler cylinder lockset provides better than average security by virtue of its internal design. Generally speaking, the more pins in a pin tumbler cylinder, the better security the lock offers. However, security also depends on the quality of the specific lock and on its application.

Quality. Quality depends on the intended service of the lockset. Light, medium, and heavy-duty locksets are available.

Type. Locksets are identified by their function (e.g., lavatory, classroom, residential, etc.)

Visual appeal. The lockset should match the decor of its surroundings; this is particularly important in new construction.

Hand. The location and direction of swing of the hinges determines the *hand* of the door (Figure 7.13). Taking the entrance side of the door as the reference point, there are four possible hands: left-hand, left-hand reverse (the door opens outward), right-hand, and right-hand reverse.

It is important to match the lockset to the door hand. Failure to do so can cause bolt/striker misalignment and result in an upside-down installation of the cylinder. When the cylinder is rotated 180°, the weight of the pins is on the springs. The pins weigh only a few grams, but that is enough to collapse the springs and disable the lock in time when a cylinder is installed upside down. Unless stipulated in the order, manufacturers supply right-hand mortise locksets; some of these can be modified in the field.

LEFT HAND: HINGES ON LEFT, OPENS INWARD.

RIGHT HAND: HINGES ON RIGHT, OPENS INWARD.

LEFT HAND REVERSE: HINGES ON LEFT, OPENS OUTWARD.

RIGHT HAND REVERSE: HINGES ON RIGHT, OPENS OUTWARD.

Figure 7.13 The *hand* of a door is a term that describes the location of the hinges and the direction of swing. (*Eaton Corp.*)

Pin Tumbler Cylinder Mortise Locks

The pin tumbler cylinder mortise lock is frequently used in homes, apartments, businesses, and large institutions. It's extremely popular and provides excellent security. Never mistake the pin tumbler cylinder mortise lock with its inferior cousin, the bit key mortise lock. Figure 7.14 shows two mortise-type pin tumbler cylinder locks without the cylinders in them.

Although the individual parts may vary slightly among different models, most basic pin tumbler cylinder mortise locks contain essentially the same parts. Practice disassembling and reassembling a few of these locks, and study the relationships of the various parts.

A — Deadbolt	**J** — Outside Hub	**T** — Latch Lever Spring
B — Deadbolt Spring	**K** — Inside Hub	**U** — Auxiliary Latch
C — Turn Knob Hub	**L** — Short Stop Button	**V** — Deadlocking Lever
D — Latch Lever	**M** — Long Stop Button	**W** — Deadlocking Lever Spring
E — Latchbolt regular	**N** — Button Cam	**X** — Stop Cylinder Lever
F — Long Hub Lever	**O** — Button Cam Spring	**Y** — Cylinder Latch Lever
G — Short Hub Lever	**P** — Latch Lever Spring	**Z** — Stop Lever
H — Hub Spring	**S** — Auxiliary Latch Lever	**AA** — Anti-Friction Latch Bolt

PARTS NOT SHOWN

BB — 1300N Hub   **DD** — 1300E Latch Lever

Figure 7.14 Inside views of pin tumbler cylinder mortise locks. (*Dominion Lock Co.*)

Figure 7.15 Kwikset's Premium Keyed Leverset. (*Kwikset Corporation*)

Specific pin tumbler cylinder locksets

The information presented in the following pages was supplied by various manufacturers. While it does not cover all pin tumbler cylinder locksets, it does give an overview of the current state of the art. If you need further information, contact a factory representative.

Kwikset Corporation. Kwikset's key-in-leversets are part of the company's Premium line (Figure 7.15). The leverset is constructed of solid brass and steel. It's engineered to meet most building code requirements, including major handicapped codes and built to absorb the punishment of high-traffic areas. It's UL listed (see Chapter 8)and backed by a 10-year limited warranty. The leverset is available in a wide range of finishes.

Kwikset offers several styles of entrance handlesets (Figure 7.16) to complement any entrance architecture from traditional to contemporary.

Kwikset's Series 500 Premium Entrance Lockset was designed and built in response to specific needs of the building industry for an improved, heavier-duty lockset for offices, apartments, townhouses, condominiums, and finer homes (Figure 7.17).

The Series 500 Locksets include features such as a panic-free operation on the interior knob; a solid steel reinforcing plate to protect the spindle assembly; a reversible deadlatch with a ⅝-inch throw, solid-brass, beveled latchbolt; a solid-steel strike plate with heavy-duty screws; and a free-spinning exterior knob for protection against wrenching.

Kwikset's Protecto-Lok (Figure 7.18) combines an entry lockset with a 1-inch cylinder deadlock. Both deadbolt and deadlatch retract with a single turn of the interior knob or lever, assuring immediate exit in case of an emergency. Protecto-Lok is UL listed when ordered and it is used with a special UL latch and strike.

Kwikset's Safe-T-Lok (Figure 7.19) combines deadbolt security with instant exit. When you lock it from the interior, the key remains in the cylinder and cannot be removed until the deadbolt is fully retracted in an unlocked position. Safe-T-Lok has a 1-inch steel deadbolt with a heat-treated steel insert that turns with any attempted cutting.

Schlage Lock Company. Schlage's B Series deadbolt locks provide primary or auxiliary security locking for both commercial and residential applications. It includes the

kwikset ENTRANCE HANDLESETS: TOLEDO DESIGN

Heavy duty

Heavy duty
with Protecto-Lok®
function

One piece

One piece
with matching
deadlock and trim

Figure 7.16 Kwikset's Entrance Handlesets. (*Kwikset Corporation*)

Figure 7.17 Kwikset's Premium Entrance Lockset. (*Kwikset Corporation*)

standard duty B100 Series for residential structures (Figure 7.20), the heavy-duty B400 Series for commercial requirements, and the most secure B500 Series for maximum performance. All products in this series feature a full 1-inch throw deadbolt with Schlage's exclusive wood frame reinforcer to deter "kick-in" attack.

Schlage's commercial quality D Series locks (Figure 7.21) are specified when the

Figure 7.18 Kwikset's Protecto-Lok. (*Kwikset Corporation*)

Figure 7.19 Kwikset's Safe-T-Lok. (*Kwikset Corporation*)

highest quality mechanisms are required. Precision manufactured to exact tolerances, the D Series is best suited for commercial, institutional, and industrial use. Specialized functions satisfy many unique locking applications.

Schlage's L Series mortise lock line is a heavy-duty commercial mortise lock series containing a wide array of knob, lever, and grip handle designs (Figure 7.22). It's available in a variety of keyed and nonkeyed functions, and in most decorative finishes. It is UL listed and can be used in offices, schools, hospitals, hotels, and commercial buildings as well as residences. The L lock meets or exceeds the ANSI A156.13 specification, making it an excellent choice for any building where security, safety, and design compatibility are prime considerations.

The L lock has one common mortise lock case for knob, lever, and grip handle trim, providing tremendous versatility. In addition, it provides excellent design flexibility in trim combinations, such as knob by lever, lever by knob, etc. All may be specified on the same L lock case.

Mounting Screws

Inside Turn Unit

2½" Adapter Ring

1" Security Insert

1" Trim Ring

Cylinder Bar

Solid Brass Pin
Tumbler Cylinder

Cylinder Housing

Wood Frame Reinforcer
with #12 × 3" wood
screws. Standard for all
B series deadbolts.

Strike Box

Strike

⅞" Diameter Housing

1" × 2¼" Self-Adjusting
Reversible Faceplate

1" Throw Deadbolt
with Concealed Hardened
Steel Roller

Door and Jamb
Preparation:
See Templates

Figure 7.20 Schlage's B Series deadbolt lock. (*Schlage Lock Company*)

The Cylinder Key

Figure 7.23 illustrates basic key nomenclature and Figure 7.24 shows an assortment of key blanks. Notice the differences in the bow, blade, and length, and in the width, number, and spacing of the grooves.

The bow usually has a specific shape that identifies the lock manufacturer. The blade length is indicative of the number of pins within the lock that the key operates (e.g., a five-pin key is shorter than a six-pin key). The height of the blade sometimes indicates the depths of the cuts in a series; generally, the higher the blade, the deeper the cuts.

Key blanks are made of brass, nickel-brass, nickel-steel, steel, aluminum, or a combination of these and other alloys. The metal determines the strength of the key and its resistance to wear.

Inside Knob

Threaded Inside Rose
holds lock firmly in place

Turn/Push Button
provides visible locking status

Long Spindle Bearing Surface
prevents wobbly knob

Friction Grip Nylon Ring
holds rose securely in place

Cylindrical Housing

**Solid Brass
6 Pin Cylinder**

Demountable Outside Knob
for quick rekeying or cylinder
replacement

Threaded Outside Rose
adjusts for door thickness

Anti-friction Latch Retractor
roller bearing & coil compression
springs for smooth knob action

**Reversible Latch Unit
½" Throw**

Deadlocking Plunger

Strike Box

Strike

**Door and Jamb
Preparation:**
See Templates

1-1/8" 2-3/4"

2-1/4"

1" dia.

2-1/8"

Figure 7.21 Schlage's D Series Key-in-knob lockset. (*Schlage Lock Company*)

Brass or Bronze Armored
Fronts

Stopworks Functions by
Turn Unit

Corrosion Protected
Steel Case and Parts

6 Pin Solid Brass Cylinder

Non-Handed Knob, Lever,
Grip Handle Trim

Stainless Steel 1" Throw
Deadbolt

Non-Handed Auxiliary
Guarded Catch

Full ¾" Latchbolt
Standard

Heavy Duty Anti-Friction
Tongue Standard

Lock Hand Catch Screw
(on lock case)

Door and Jamb
Preparation:
See Templates

Armor Front Adjustable
for Door Bevel

Fusible Link Built into
Lock Case

Figure 7.22 Schlage's L Series mortise lockset. (*Schlage Lock Company*)

Figure 7.23 Basic cylindrical key nomenclature.

Figure 7.24 Assorted key blanks. (*Master Lock Company*)

Duplicating a cylinder key by hand

Duplicating a cylinder key by hand isn't much different than duplicating a lever or warded key by hand. The procedure is as follows:

1. Smoke the original and mount it together with the blank in a vise. The blank must be in perfect alignment with the original.

2. Using Swiss round and warding files, begin at the tip of the key and work toward the bow. Start in the center of each cut.

3. Go slowly near the end of each cut, but continue to cut with firm and steady strokes. Stop when the file touches the soot on the original key.

4. Once you have finished, remove the keys from the vise. Wire-brush the duplicate to remove any burrs, then polish the key.

5. Hold the duplicate up to the light. Place the original in front of it and behind it to determine if the duplicate is accurately cut. Shallow cuts can be cut a little deeper with a file; if a cut is too deep, start over with a new key blank.

Impressioning

The technique for impressioning cylinder keys is different than for warded keys. You cannot smoke the cylinder key because the soot would wipe off the blade when you insert the key into a pin tumbler cylinder.

Instead, depend on small marks left on the key blank by the pin tumblers. These marks indicate the position of the tumblers relative to the blank and the depth of each cut. Proceed as follows:

1. Examine the blank. The marks left by the tumblers are difficult to see in the best of circumstances; scratches across the top of the blade will make them impossible to see. Polish the top of the blade with emery paper.

2. Place the bow of the key in a small C-clamp, and insert the blade into the keyway.

3. Twist the blank side to side and gently rock it up and down; there's no need to twist very hard. If you break the key blank, you've used far too much pressure.

4. The tumblers should leave their "footprints" on the blade. Examine the blank under a bright light.

5. Identify each mark with a light file cut.

6. Using a Swiss No. 4 file, make a shallow cut at the mark closest to the tip of the blank.

7. Insert the blank into the cylinder, twist the blank from side to side, gently rock it up and down, then remove it. Check the mark and file a little more. Continue until the tumbler no longer leaves a mark.

When you finish one cut, move to the next and repeat the procedure. Keep the slopes even and smooth.

When the last cut is finished, the newly cut key should operate the cylinder. Smooth the various angles on the key and remove any burrs.

Vending machine locks

The tubular key lock has become standard on vending machines. The arrangement of the pin tumblers increases security and requires a different type of key—one that is tubular. A typical tubular key lock and key is shown in Figure 7.25. The cam works directly off the end of the plug.

The key has its bitting disposed radially on its end. The depth and spacing of each cut must match the pin arrangement. In addition, there are two notches, one on the inside and the other on the outer edge of the key. These notches align the key to the lock face. Otherwise the key could enter at any position.

The key bittings push the pins back, bringing them to the shear line. Once the pins are in alignment the key is free to turn the plug and attached cam.

The pins within the lock are entirely conventional in construction, with the exception of the bottom one. A ball bearing is sandwiched between the pin and its driver. The bearing reduces friction and increases pin life. The pin in question is not interchangeable with others in the lock. Pin tolerance is extremely critical. There is no room for sloppy key cutting.

Figure 7.25 An "Ace" Tubular Key Lock.

Disassembly. To disassemble the tubular key lock, follow this procedure:

1. Place the lock into its holder. Ace makes a special vise for these cylinders.

2. Drill out the retainer pin at the top of the assembly. Use a No. 29 drill bit and stop before the bit bottoms in the hole (Figure 7.26).

NO. 29 DRILL

RETAINING PIN

Figure 7.26 Use a No. 29 drill bit to remove the retaining pin. Alternately, use a No. 42 bit, thread a small metal screw into the hole, and extract the pin and screw by prying upward on the screw head. (*Desert Publications*)

3. Remove what is left of the pin with a screw extractor.

4. Insert the appropriate plug follower into the cylinder. Apply light pressure. Plug follower dimensions are: length 1.50 inches; outside diameter 0.373 inch; inside diameter 0.312 inch.

5. Lift off the outer casing from the bushing assembly.

6. Scribe reference marks on the plug sections as an assembly guide.

7. Remove the follower from the cylinder. *Note:* Perform this operation carefully. The pins are under spring tension and must be kept in order. If not, you will have a monumental job sorting the pins. There are 823,543 possible combinations!

Assembly.   Assemble the lock in the reverse order of disassembly. Replace the retaining pin with an Ace part, available from locksmith supply houses.

Rekeying.   Rekeying is not difficult if you approach the job in an orderly manner.

1. Cut the key. There are seven bit depths, ranging from 0.020 inch to 0.110 inch in 0.015-inch increments.

2. Remove the pins with tweezers. *Note:* Pin lengths range from 0.025 inch to 0.295 inch in increments of 0.015 inch. Drivers are available in 0.125-inch, 0.140-inch, and 0.180-inch lengths.

3. Select new pins using the key combination as a guide.

4. Install the pins. The flat ends of the core pins are toward the key (Figure 7.25). This pattern must be followed:

   Core pins 1, 2, and 3 require 0.180-inch drivers.
   Core pins 4 and 5 require 0.140-inch drivers.
   Core pins 6 and 7 require 0.125-inch drivers.
   Note: Pins are numbered clockwise from the top as you face the lock. Insert the key.

5. Install the plug in the cylinder with the scribe marks aligned.

6. Insert a new retaining pin and give it a sharp tap with a punch.

7. Test the key. It may be necessary to rap the cam end of the plug with a mallet.

Lockout

A lockout can be a real headache. Ordinary tubular key locks—those without ball bearings—can be drilled out with the tool shown in Figure 7.27. Hole saws are available for standard and oversized keyways.

Figure 7.27 Drilling the lock requires a piloted hole saw, available from locksmith supply houses. (*Desert Publications*)

High-Security Mechanical Locks

With respect to locking devices, the term *high-security* has no precise meaning. Some manufacturers take advantage of that fact by arbitrarily using the term to promote their locks.

Most locksmiths would agree that to be considered *high-security* a lock should have features that offer more than ordinary resistance against picking, impressioning, drilling, wrenching, and other common forms of burglary. The most secure locks also provide a high level of key control. The harder it is for an unauthorized person to have a duplicate key made, the more security the lock provides.

Underwriters Laboratories Listing

Founded in 1894, Underwriters Laboratories, Inc. (UL) is an independent, nonprofit product-testing organization. A UL listing (based on UL standard 437) is a good indication that a door lock or cylinder offers a high degree of security. If a lock or cylinder has such a listing, you'll see the UL symbol on its packaging or on the face of the cylinder.

To earn the UL listing, a lock or cylinder must meet strict construction guidelines and a sample model must pass rigorous performance and attack tests. Some of the requirements are as follows:

- All working parts of the mechanism must be constructed of brass, bronze, stainless steel, or equivalent corrosion-resistant materials or have a protective finish complying with UL's Salt Spray Corrosion test.

- Have at least 1000 key changes.

- Operate as intended during 10,000 complete cycles of operation at a rate not exceeding 50 cycles per minute.

- The lock must not open or be compromised as a result of attack tests using hammers, chisels, screwdrivers, jaw-gripping wrenches, pliers, hand-held electric drills, saws, puller mechanisms, key impressioning tools, and picking tools.

The attack test includes 10 minutes of picking, 10 minutes of key impressioning, 5 minutes of forcing, 5 minutes of drilling, 5 minutes of sawing, 5 minutes of prying, 5 minutes of pulling, and 5 minutes of driving. Those are net working times, which don't include time used for inserting drill bits or otherwise preparing tools.

Key Control

Another important factor in lock security is *key control*. The most secure locks have patented key blanks that are available only from the lock manufacturer. At the next level of key control are key blanks that can be cut only on special key machines. This type of key control greatly reduces the number of places where an unauthorized person can have a key duplicated. The least secure locks use keys that can be copied at virtually any hardware or department store.

Key patents

Not all patented keys offer a lot of key control. There are two types of patents: design and utility. A *design patent* only protects how a thing looks; a *utility patent* protects the way a thing works. In 1935, Walter Schlage received a design patent for a key bow. The bow's distinctive shape made it easy for locksmiths and other key cutters to recognize the company's keys. However, the patent didn't prevent aftermarket key blank manufacturers from making basically the same key blanks with slightly different bow designs. Utility patents provide more protection against unauthorized manufacturing of key blanks.

All patents eventually expire, and they aren't renewable. Design patents are granted for 14 years. Utility patents are granted for 17 years. Medeco Security Lock's original utility patent (No. 3,499,302) was issued in March 1970. Because that patent prevented companies from making key blanks that fit Medeco locks, the company was able to maintain maximum key control. As soon as the patent expired (in 1987), compatible key blanks were offered by many companies. The company responded by introducing a new lock, the Medeco Biaxial, and getting a utility patent (No. 4,635,455) for it in June 1987. That should protect the key blank until June 2004.

Other important utility patents include: Assa (No. 4,393,673), granted July 1983; Schlage Primus (No. 4,756,177), granted July 1988; Kaba Gemini (No. 4,823,575), granted April 1989; and Kaba Peaks (No. 5,016,455), granted May 1991.

If key control is important to your customer, make sure you find out what type of patent a lock has and when the patent was issued.

Types of High-Security Mechanical Locks

This section provides detailed information about some of the most popular high-security mechanical locks available. Much of the information was obtained from manufacturer's technical bulletins and service manuals.

Schlage's Primus

The Schlage Primus is one of Schlage Lock Company's newest high-security mechanical locking systems. It features a specially designed, patent protected key that operates either the UL listed #20-500 High Security Series Cylinders or the Controlled Access #20-700 Series Cylinders (not UL listed). Both series are available in Schlage A, B, C/D, E, H, and L Series locks (Figures 8.1 and 8.2).

Both series are easily retrofitted into existing Schlage locks, and they can be keyed into the same master key system and operated by a single Primus key. The

Figure 8.1 An exploded view of a Schlage Primus mortise cylinder. (*Schlage*)

Figure 8.2 An exploded view of a Schlage Primus Key-in-knob cylinder. (*Schlage*)

Primus key is cut to operate all cylinders while those that operate the standard cylinders will not enter a Primus keyway.

The Primus security cylinder is machined to accept a side bar and a set of five finger pins, which, in combination with Schlage's conventional six pin keying system, provides two independent locking principles operated by a Primus key. Hardened steel pins are incorporated into the cylinder plug and housing to resist drilling attempts.

Primus security levels.   The Primus system features five different levels of security. Each level requires an appropriate ID card for key duplication.

Security Level One.   Primus cylinders and keys have a standard side milling allocated to Level One and are on a local stock basis. Level One also provides the flexibility of local keying into most existing Schlage key sections; it utilizes a Primus key to operate both systems. Key control is in the hands of the owner, who holds an ID card for the purpose of acquiring additional keys. Level One is serviced through qualified locksmiths who are fully trained at a Schlage Primus I Center.

Security Level One Plus.   Level One Plus increases Level One security by allocating a restricted side-bit milling to a limited number of local Primus centers, which also control keying and key records.

Security Level Two.   At Level Two, service is carried out solely by original Schlage Primus distributors. Also, key duplication for the exclusive side-bit milling requires the authorized purchaser's signature. Factory keying is also available.

Security Level Three.   At Level Three, the Schlage Lock Company maintains control of the inventory, keying, keying records, installation data, and owner's signature. Keys for the geographically restricted, side-bit millings can be duplicated only by the factory after the owner's signature has been verified.

Security Level Four.   Level Four provides all the restrictions of Level Three, plus an exclusive side-bit milling.

Kaba

Kaba locks are *dimple key* locks. There isn't just one, but an entire family of different Kaba cylinder designs. Each one is designed to fill specific security requirements (Figures 8.3, 8.4, and 8.5).

The family includes Kaba 8, Kaba 14, Kaba 20, Kaba 20S, Saturn, Gemini, and Micro. Those which use numbers in their designations generally reflect not only the order of their development, but also the number of possible tumbler locations in a cylinder of that particular design (Figures 8.6 and 8.7).

Handing and key reading.   The concept of *handing* is basic to the understanding of all Kaba cylinders. With Kaba cylinders handing is not a functional installation limitation as you might expect. A left-hand cylinder will operate both clockwise and counterclockwise and function properly in a lock of any hand. The handing of Kaba cylinders refers only to the positions where the pin chambers are drilled.

Figure 8.3 Some Kaba locks fit Adams Rite narrow stile metal doors. (*Lori Corp.*)

Figure 8.4 Double cylinder deadbolt lock with "Inner Sanctum" core. (*Lori Corp.*)

Figure 8.5 Micro-Kaba switch lock and keys. On the right is the lock with the core removed. The switch size (compare with coin) provides for a wide variety of uses with electrical and electronic circuitry. (*Lori Corp.*)

Figure 8.6 Kaba standard mortise cylinders can be used on Adams Rite doors. (*Lori Corp.*)

Figure 8.7 Cutaway view of a Kaba 20 cylinder. (*Lori Corp.*)

Figure 8.8 shows the orientation of the two rows of side pins in a Kaba 14 cylinder. Notice that they are staggered much like the disc tumblers in some foreign car lock cylinders.

There are two possible orientations of these staggered rows of side pins. Either row could start closer to the front of the cylinder. The opposite row would then start farther from the front. These two orientations are referred to as *right-* or *left-hand*. If the cylinder is viewed with the right side up, the hand of most Kaba designs is determined by the side whose row of pins begins farther from the front when viewing the face of the cylinder.

Figure 8.8 Kaba keys are handed. (*Lori Corp.*)

If a key is to do its job and operate a cylinder, obviously the cuts must be in the same positions as the pins in the cylinder. That means there are two ways to drill dimples in the keys as well.

To determine the hand of a Kaba key, view it as though it was hanging on your key board (Figure 8.9). Notice that one row of cuts starts farther from the bow than the other. The row that starts farther from the bow determines the hand. If the row starting farther from the bow is on the left side, it is a left-hand key. If the row starting farther from the bow is on the right side, it is a right-hand key.

Occasionally, you may find a Kaba key with both right-hand and left-hand cuts (Figure 8.10). This key is called *composite bitted*. It is primarily used in maison key systems (a type of master key system).

Key reading.   Determining the hand is the first step in key reading. Next, you need to know the order in which the various dimple positions are read.

The positions are always read bow to tip beginning with the hand side. After the hand side comes the nonhand side, then the edge. Remember to go back to the bow

Figure 8.9 It's easy to determine the hand of a Kaba key. (*Lori Corp.*)

Figure 8.10 Examples of Kaba composite bitted keys. (*Lori Corp.*)

Composite Bitted Keys

to start each row (Figure 8.11). For composite bitted keys, use a Kaba key gauge to help mark the cuts of both hands.

All Kaba designs except Micro have four depths on the sides; Micro has three. Kaba's depths are numbered opposite from the way we would normally expect. Number 1 is the deepest and #4 is the shallowest. These depths can be read by eye with very little practice (Figure 8.12). The increment for side depths of Kaba 8 and 14 is 0.0157 inch (0.4 mm). For Gemini, it is 0.0138 inch (0.35 mm).

For the edges, reading is a bit different for the various Kaba designs. The Kaba key gauge is very useful for reading the edges. Find the section of the gauge which corresponds to the design of the Kaba key you're attempting to read (e.g., Kaba 8). Place the key under the gauge so it shows up through the edge slot. When the shoulder of the key hits the stop on the gauge, you're ready to read the positions of the cuts.

Because of the nonstandard cylinder drilling for Kaba 8, the codes show two columns for the edge (Figure 8.13). There are only two depth possibilities on the edge of Kaba 8 and Kaba 14: *cut = #2* and *no-cut = #4*. The combinations in Figure 8.13 call for a key with a cut only in position 4, while the cylinder is drilled in posi-

Figure 8.11 Positions on Kaba keys. (*Lori Corp.*)

Figure 8.12 Depths on Kaba keys. (*Lori Corp.*)

S I D E S		E D G E	
		key	cyl.
R2214	4233	4	147
R2242	4233	4	147
R2412	4233	4	147
R4212	4233	4	147
R2244	1233	4	147
R2414	1233	4	147

Figure 8.13 Kaba 8 combinations. (*Lori Corp.*)

tions 1, 4, and 7. This means the cylinder receives a #2 pin in position 4 for the key cut. Positions 1 and 7 require a #4 pin because there is no cut on the key.

For the Kaba 14 and new Kaba 8 using ME series codes, the edge is read differently because you know automatically which positions are involved in every case. These cylinders are all drilled with odd edges (positions 1, 3, 5, and 7).

Knowing the positions involved, the edge combination of these keys is notated in terms of depths, rather than positions. A Kaba key gauge can be used to determine the positions of the #2 edge cuts. If there is a cut in position 5 only, the combination would be 4224. If there are cuts in positions 1, 3 and 7, the combination would be 2242. If there are cuts in positions 3 and 5, the combination would be 4224, etc. We know there are four chambers in the cylinder and they are in the odd-numbered positions. Therefore, we have to come up with four bittings in the edge combination. *Remember:* cut = #2 and no-cut = #4.

For Kaba Gemini, there are three active depths on the edge, plus a high #4 cut used in masterkeying. Because there is no *no-cut* on Gemini, a key gauge should not be necessary to read the edge combination. Right-hand stock keys will always have cuts in positions 3, 5, 7, and 9 and left-hand keys for factory master key systems will almost always have cuts in positions 2, 4, 6, 8, and 10 (Figure 8.14).

Bitting notation. Before the bitting comes an indication for the hand, "R" or "L." As mentioned, all bittings are read and notated, bow to tip. The key combination is broken up into separate parts corresponding to each row of pins in the cylinder. The first group of bittings is the hand side, the second group is the opposite side, and the third is the edge (Figure 8.15).

This holds true for all Kaba designs, but the guard pin cut on a Saturn key (a #3 depth) is not part of the key combination and should be ignored for all phases of key reading.

Notation of composite bitted key combinations isn't much different from that of regular keys. Composite bitted keys have both right- and left-hand side cuts and often have both even and odd position edge cuts. Such keys are normally used only in selective key systems and maison key systems.

If most of the key system is made up of left-hand cylinders, that is the hand which is listed first. Conventionally, the edge bittings are all listed as part of the first line. The opposite hand bittings are written under those of the main hand, as illustrated in Figure 8.16.

STOCK   MK'd
3-5-79   2-4-6-8-10

Figure 8.14 Kaba Gemini edge. (*Lori Corp.*)

Kaba 8   R 1421-2414-17
Kaba 14   L 34121-14123-2442
Kaba Gemini (Stock)   R 14214-241414-1231
Kaba Gemini (MK'd)   L 11314-442121-14321
Kaba Saturn   L 142-313

Figure 8.15 Writing key combinations. (*Lori Corp.*)

L 43121-124231-314213112
R 24133-122431

Figure 8.16 Composite bitted key. (*Lori Corp.*)

Identifying nonoriginal keys. Because the dimensions of the key blanks for K-8, K-14, and Saturn are identical, people in the field sometimes duplicate a key from one design onto another's key blank. This can lead to problems later, especially when quoting prices to a customer. If a key says "Kaba 8" but it has really been cut for a Kaba 14 cylinder, you might quote a Kaba 8 price and order a Kaba 8 cylinder only to find that you can't set it to the customer's key. To avoid confusion at a later date, a genuine key blank with a system designation such as "Kaba 8," "Kaba 14," or "Kaba Saturn" should only be used to make keys for that particular design. If you didn't sell the job originally, you should always verify the design by counting the dimples on the key.

If you need a new cylinder to match a nonoriginal key or if the key was made poorly and you must make a code original to operate the lock properly, you must also be able to determine which Kaba design the key was cut for. This is easily done by counting the dimples on the sides, being careful not to overlook any of the tiny #4 cuts.

Kaba 8 has two rows of four cuts (eight total), as seen in Figure 8.17. Kaba 14 has two rows of five cuts (10 total) (Figure 8.18.) Saturn has a row of three and a row of four (seven total). Kaba Gemini uses a different key blank that is thicker and narrower. The dimples are oblong rather than round (Figure 8.19). It has a row of five cuts and a row of six cuts (11 total).

Composite bitted keys have exactly twice as many dimples on a side.

Figure 8.17 Eight bittings equal Kaba 8. (*Lori Corp.*)

Figure 8.18 Ten bittings equal Kaba 14. (*Lori Corp.*)

Figure 8.19 Eleven oblong bittings equal Kaba Gemini. (*Lori Corp.*)

Medeco locks

Locks manufactured by Medeco Security Locks, Inc. are perhaps the most well known high-security mechanical locks in North America. For that reason, much of this remaining section is devoted to reviewing how the various types of Medeco locks operate.

General operation.   Before studying in detail the specifications of Medeco locks, you must first understand how these locks operate. Medeco's 10- through 50-series locks incorporate the basic principles of a standard pin tumbler cylinder lock mechanism. A plug, rotating within a shell, turns a tailpiece or cam when pins of various lengths are aligned at a shear line by a key. Figure 8.20 shows an exploded view of a Medeco cylinder.

The plug rotation in a Medeco lock is blocked by the secondary locking action of a sidebar protruding into the shell. Pins in a Medeco lock have a slot along one side. The pins must be rotated so that this slot aligns with the legs of the sidebar. The tips of the bottom pins in a Medeco lock are chisel pointed, and they are rotated by the action of the tumbler spring seating them on the corresponding angle cuts on a Medeco key (Figure 8.21).

The pins must be elevated to the shear line and rotated to the correct angle simultaneously before the plug will turn within the shell (Figure 8.22). This dual-locking principle and the cylinder's exacting tolerances account for Medeco's extreme pick resistance.

Medeco cylinders are also protected by hardened, drill-resistant inserts against other forms of physical attack. There are two hardened, crescent-shaped plates within the shell that protect the shear line and the side bar from drilling attempts. There are also hardened rods within the face of the plug and a ball bearing in the front of the sidebar. Figure 8.23 shows the shapes and positions of the inserts.

To fit within the smaller dimensions necessary in a cam lock, Medeco developed the principle of a *driverless rotating pin tumbler*. It is used in the 60- through 65-series locks. The tumbler pin and springs are completely contained within the plug diameter (Figure 8.24).

Figure 8.20 Exploded view of an original Medeco mortise cylinder. (*Medeco Security Locks*)

ASSEMBLY NUMBER 10-200 - MORTISE CYLINDER			
ITEM	DESCRIPTION	REQ'D	PART NO.
1	Key 6 Pin	1	10-011
2	Plug Mortise 6 Pin	1	10-012
3	Shell Mortise 6 Pin	1	10-017
4	Fence 6 Pin	1	10-022
5	Washer	1	10-025
6	Cam	1	10-027
7	Spring Cover 6 Pin	1	10-030
8	Screw #2-56 Slotted F H	2	10-032
9	Spring "Tumbler"	6	10-034
10	Driver (Hardened on application)	6	10-035 - 10-040
11	Tumbler	6	Listed below
12	Spring Fence	2	10-047
13	Master Pin Disc	As required	10-048 - 10-052

Figure 8.21 Angle cuts on a Medeco key cause the tumblers in a Medeco cylinder to be raised to the shear line while simultaneously rotating into position to allow the sidebar's legs to push into the pins. (*Medeco Security Locks*)

LOCKING SLOT
IN CYLINDER

SIDEBAR
IN LOCKED POSITION

Figure 8.22 A cutaway view of an original Medeco mortise cylinder showing how both the pins and sidebar obstruct the plug from rotating. (*Medeco Security Locks*)

Figure 8.23 Face of a Medeco cylinder showing the positions of the hardened steel drill resistant pins and crescents. (*Medeco Security Locks*)

The plug rotation is blocked by the locking action of a sidebar protruding into the shell. The pins are chisel pointed and have a small hole drilled into the side of them. The pins must be rotated and elevated by corresponding angled cuts on the key so that each hole aligns with a leg of the sidebar and allows the plug to rotate. In addition, the cylinder is protected against other forms of physical attack by four hardened, drill-resistant rods within the face of the plug.

Despite the exacting tolerances and additional parts, Medeco's cylinders are less susceptible to wear problems than are conventional cylinders. As in all standard pin tumbler cylinder locks, the tips of the pins and the ridges formed by the adjacent cuts on the key wear from repeated key insertion and removal.

Figure 8.24 Exploded view of an original Medeco cam lock. (*Medeco Security Locks*)

In a Medeco lock, this wear has little effect on its operation because, in contrast to a standard lock cylinder, the tips of the pins in a Medeco lock never contact the flat bottoms of the key cuts. Instead, they rest on the sides of the key profile; thus, the wear on the tips of the pins does not affect the cylinder's operation (Figure 8.25). Cycle tests in excess of one million operations have proven Medeco's superior wear resistance.

Figure 8.25 Angles of an original Medeco bottom pin and key cut. (*Medeco Security Locks*)

Medeco keys. There are four dimensional specifications for each cut on a Medeco key. They are the cut profile, the cut spacing, the cut depth, and the cut angular rotation.

The profile of the cut on all Medeco keys must maintain an 86° angle. This dimension is critical because the pins in a Medeco lock are chisel pointed and seat on the sides of the cut profile rather than at the bottom of the cut. Prior to June 1975, Medeco keys were manufactured with a perfect V-shaped profile. After this date, the keys were manufactured with a 0.015-inch-wide flat at the bottom.

Spacing of the cut on a Medeco key must be to manufacturers' specifications. For the 10-series stock keys, Medeco part KY-105600-0000 (old #10-010-0000) and KY-106600-0000 (old #10-011-0000), the distance from the upper and lower shoulder to the center of the first cut is 0.244 inch. Subsequent cuts are centered an additional 0.170 inch. For the 60-series stock keys, Medeco part KY-105400-60000 (old #60-010-6000) (five pin) and KY-104400-6000 (old #60-011-6000) (four pin), the distance from the upper shoulder to the center of the first cut is 0.216 inch. Subsequent cuts are centered an additional 0.170 inch. The distance from the bottom shoulder to the center of the first cut is 0.244 inch on this and all Medeco keys. For the 60-series thick head keys, Medeco part KY-114400-6000 (old #60-611-6000) (five pin) and KY-114400-6000 (old #60-611-6000) (four pin), the distance from the shoulder to the center of the first cut is 0.244 inch. Subsequent cuts are centered an additional 0.170 inch.

Standard Medeco keys in the 10- through 50-series are cut to six levels with a full 0.030-inch increment in depths. Keys used in extensive masterkeyed systems and on restricted Omega keyways are cut to eleven levels with a half step 0.015-inch increment in depths.

Because of the size limitations, Medeco keys in the 60- through 65-series are cut to four levels with a 0.030-inch increment in depth. Keys used in extensive masterkeyed systems and Omega keyways are cut to seven levels with a 0.015-inch increment in depth.

In addition to the dimensions above, each cut in a Medeco key may be cut with any one of three angular rotations. These rotations are designated as right (R), left (L), or center (C). When you look into the cuts of a Medeco key as illustrated in Figure 8.26, concentrating on the flats of the cut, you will notice that flats that are positioned perpendicular to the blade of the key are designated *center angles*. Flats that point upward to the right are designated *right angles*, and flats that point upward to the left are designated *left angles*. Right and left angles are cut on an axis 20° from perpendicular to the blade of the key.

Keyways and key blanks. The entire line of Medeco locks is available in numerous keyways. The use and distribution of key blanks of various keyways is part of Medeco's systematic approach to key control. Medeco offers four levels of security:

Signature Program. Consumers can get new keys made simply by going to the locksmith from whom the lock was purchased. Keys for the Signature Program are restricted by Medeco to locksmiths who have contracted with the company.

Figure 8.26 Angles of cuts in an original Medeco key.

Card Program. A consumer is issued a card embossed with control data and that has space for a signature. The keys are owned by Medeco and are controlled by contract. An authorized locksmith will make duplicate keys only after verifying the card data and signature.

Contract restricted. A business or institution can enter into a contract with Medeco for a specially assigned keyway. The key blanks can then be ordered through any authorized Medeco distributor with appropriate authorization of the business or institution.

Factory Program. The Factory Program is Medeco's highest level of key control for consumers, businesses, and institutions. Keys are made from factory restricted key blanks that are never sold by the factory. Duplicate keys are available only directly from the factory on receipt and verification of an authorized signature.

10-series pins. Medeco bottom pins differ significantly from standard cylinder pins in four respects. The differences occur in the diameter, the chisel point, the locator tab, and the sidebar slot.

- Medeco pins have a diameter of 0.135 inch. That is 0.020 inch larger than the 0.115-inch diameter pin in a standard pin tumbler cylinder.
- All bottom pins are chisel pointed with an 85° angle. The tip is also blunted and beveled to allow smooth key insertion.
- The *locator tab* is a minute projection at the top end of the tumbler pin. The locator tab is confined in a broaching in the shell and the plug; it prevents the bottom pin from rotating a full 180°. A 180° rotation causes a lockout because the sidebar leg is not able to enter the sidebar slot in the pin.
- The *sidebar slot* is a longitudinal groove milled in the side of the bottom pins to receive the sidebar leg. The slot is milled at one of three locations in relationship to the axis of the chisel point.

In 1986 the patent for the original Medeco lock system expired. Now other key blank manufacturers can produce those key blanks. The original system, referred to by Medeco as "controlled," is still being used throughout the world. To allow greater key control, however, Medeco produced a new lock system called the Medeco Biaxial System.

Medeco's Biaxial lock system. From the outside, Medeco Biaxial cylinders look similar to those of the original system. Internally, however, there are some significant differences. Figures 8.27 and 8.28 show exploded views of Medeco Biaxial cylinders.

Assembly Number 10-0200 Mortise Cylinder			
Item	Description	Req'd.	Part No.
1	Key 6 Pin	1	KY-176600-xxxx
2	Plug Mortise 6 Pin	1	12-005006-xx
3	Shell Mortise 6 Pin	1	13-009267
4	Sidebar 6 Pin	1	CP-084621
5	Cam Washer	1	CP-021011
6	Cam	1	as required
7	Set Screws	6	CP-011013-020
8	Screw #2-56 Slotted FH	2	CP-010103-050
9	Spring	6	CP-030001
10	Top Pin	6	10 Series
11	Bottom Pin	6	50 Series
12	Sidebar Spring	2	CP-030018
13	Master Pin (Wafer)	As required	

**Certain mortise and rim top pins are hardened steel to prevent drilling of shear line.

Assembly Number 10-0400 Rim Cylinder			
Item	Description	Req'd.	Part No.
1	Key 6 Pin	1	KY-176600-xxxx
2A	Plug Rim 6 Pin	1	12-032006
3A	Shell Rim 6 Pin	1	13-015267
4	Sidebar 6 Pin	1	CP-084621
5A	Plug Retainer	1	CP-020010
6A	Tailpiece	1	As required
7	Set Screws	1	CP-011013-020
9	Spring	6	CP-030001
10	Top Pin	6	10 Series
11	Bottom Pin	6	50 Series
12	Sidebar Spring	2	CP-030018
13	Master Pin (Wafer)	As required	
14	Mounting Screws	2	CP-011417-000
*15	Steel Insert (Hardened)	2	CP-180011
*16	Steel Ball (Hardened) pre-assy. w/item #4	1	CP-250060
*17	Cylinder Collar	As required	CP-180021
*18	Security Pin pre-assy. w/item #2	2	CP-060103
*19	Security Pin pre-assy. w/item #2	1	CP-060100
20	Mounting Plate	1	CP-180031
*21	Mushroom Drivers	As required	10 Series
22	Tailpiece Retainer	1	CP-021121
23	Roll Pin	1	CP-060203

*These parts included on 10-0200 MORTISE CYLINDER
Patent No. 3499302

Figure 8.27 Exploded view of a Medeco biaxial mortise and rim cylinder. (*Medeco Security Locks*)

Figure 8.28 Exploded view of a Medeco biaxial key-in-knob cylinder. (*Medeco Security Locks*)

1. Plug
2. Sidebar
3. Sidebar Springs
4. Bottom Pins
5. Top Pins
6. Springs
7. Spring Cover
8. Shell
9. Security Pins
10. Tailpiece
11. Retainer
12. Retainer Screws
13. Key

Biaxial pins differ from original Medeco pins in three respects: the chisel point, the locator tab, and the pin length. Biaxial pins are made of CDA340 hard brass and are electroless nickel plated. They have a diameter of 0.135 inch and are chisel pointed with an 85° angle. However, this chisel point is offset 0.031 inch in front of or to the rear of the true axis or center line of the pin (Figure 8.29).

Fore pins are available in three angles: B, K and Q. Aft pins are also available in three angles: D, M, and S. Pins B and D have a slot milled directly above the true centerline of the pins. Pins K and M have a slot milled 20° to the left of the true centerline of the pin. Pins Q and S have slots milled 20° to the right of the true centerline of the pin (Figure 8.30).

Fore Pin   Aft Pin

Figure 8.29 Medeco biaxial pins. (*Medeco Security Locks*)

Figure 8.30 Angular rotations of a Medeco biaxial key. (*Medeco Security Locks*)

The locator tab, the minute projection limiting the pin rotation, was moved to the side of the pin roughly 90°. It is now located along the centerline of the pin opposite the area for sidebar slot (Figure 8.31).

Medeco Biaxial key specifications. There are four dimensional specifications for a Medeco Biaxial key: the cut profile, the cut depth, the cut spacing, and the cut angular rotation.

The cut profile of the Medeco Biaxial key remains at 86°. However, keys in the 10- through 50-series locks are cut to six levels with a full 0.025-inch increment in depths.

Spacing of the cut on Medeco Biaxial keys must be to manufacturers' specifications. Because Medeco Biaxial pins have the chisel point forward or aft of the pin centerline, the dimensional spacing on the Biaxial key blank can change from chamber to chamber.

From the shoulder to the center of the first cut using a fore pin (either B, K, or Q), the dimension will be 0.213 inch. From the shoulder to the center of the first cut using an aft pin (either a D, M, or S), the dimension will be a 0.275 inch. Subsequent fore cuts and subsequent aft cuts are spaced at 0.170 inch (Figure 8.32).

While there are only three angular rotations on a key, each rotation can be used with either a fore pin or an aft pin. Angular cuts B and D are perpendicular to the blade of the key; B is a fore cut and D is an aft cut. Angular cuts K and M have flats pointing upward to the left; K is a fore cut and M is an aft cut. Angular cuts Q and S have flats pointing upward to the right; Q is a fore cut and S is an aft cut. K, M, Q, and S angles are cut on an axis 20° from perpendicular to the blade of the key (Figure 8.33).

Figure 8.31 The locator tab is located along the centerline of the Medeco biaxial pin opposite the area for sidebar slot. (*Medeco Security Locks*)

Figure 8.32 Spacing for Medeco biaxial keys. (*Medeco Security Locks*)

Figure 8.33 Angles for cuts in Medeco biaxial keys. (*Medeco Security Locks*)

9

Masterkeying

Masterkeying can provide immediate and long range benefits that a beginning lock-smith can find most desirable. This chapter covers the principles of masterkeying and techniques for developing master key systems.

Coding Systems

Coding systems help the locksmith distinguish various key cuts and tumbler arrange-ments. Without coding systems, masterkeying would be nearly impossible.

Most coding systems (those for disc, pin, and lever tumbler locks) are based on depth differentiation. Each key cut is coded according to its depth; likewise, each matching tumbler receives the same code. Depths for key cuts and tumblers are standardized for two reasons: it is more economical to standardize these depths (mass production would be impossible without some kind of standardization); and, depths, to some extent, are determined by production.

Master Key Systems

In most key coding systems, tumblers can be set to any of five possible depths. These depths are usually numbered consecutively 1 through 5. Since most locks have five tumblers, each one with five possible settings, there can be thousands of combina-tions. Master keys are possible because a single key can be cut to match several lock combinations.

In developing codes, there are certain undesirable combinations which cannot be used. The variation in depths between adjoining tumblers cannot be too great. For ex-ample, a pin tumbler key cannot be cut to the combination 21919 because the cuts for the 9s rule out the cuts for the 1s. Likewise, a pin tumbler lock with the combination 99999 would be too easy to pick. The undesirable code combinations vary depending on the type of tumblers, the coding system, and the number of possible key variations. The more complex the system, the greater the possibility of undesirable combinations.

Masterkeying Warded Locks

Since a ward is an obstruction within a lock that keeps out certain keys not designed for the lock, a master key for warded locks must be capable of bypassing the wards. Figure 9.1 shows a variety of side ward cuts that are possible on warded keys. The master key (marked *M*) is cut to bypass all the wards in a lock admitting the other six keys.

As explained earlier, cuts are also made along the length of the bit of a warded key. These cuts correspond to wards in the lock. To bypass such wards, a master key must be narrowed.

Because of the limited spaces on a warded key, masterkeying is limited in the warded lock. The warded lock, because it offers only a very limited degree of security, uses only the simplest of master key systems. Figure 9.2 shows some of the standard master keys that are available from factories.

Figure 9.1 The master key at the bottom of the drawing replaces the six change keys above it.

Figure 9.2 Factory-supplied warded master keys.

Figure 9.3 Double gated lever cuts (shown by dotted lines) should be avoided.

Figure 9.4 Wide gated cuts are preferred.

Masterkeying Lever Tumbler Locks

Individual lever locks may be masterkeyed locally, but any system that requires a wide division of keys would have to be set up at the factory. A large selection of tumblers is required. The time involved in assembling a large number would make the job prohibitive for the average locksmith.

There will be occasions when you are asked to masterkey small lever locks. There are two systems. The first is the *double-gate* system (Figure 9.3); the other is the *wide-gate* system (Figure 9.4). Double gating is insecure. As the number of gates in the system increases, care must be taken to prevent cross-operation between the change keys. For example, you may find a change key for one lock acting as the master key.

With either system, begin by determining the tumbler variations for the lock series in question. If the keys to all the locks are available, read the numbers stamped on the keys. Otherwise, disassemble the locks and note the tumbler depths for each one. Next, make a chart listing the tumbler variations (Figure 9.5).

The master key combination can be set up fairly easily now. The chart in Figure 9.5 is for ten lever locks, each with five levers with five possible key depths per lever. The master key for these locks will have a 21244 cutting code.

Suppose the tumblers in the first position have depths of 1, 2, 3, and 5. You must file depths 1 and 3 wider to allow the depth 2 cut of the master key to enter. The tumbler with the depth 5 cut requires a separate cut, or double gating. It needs a cut that will align it properly at two positions.

TUMBLER POSITIONS

	POSITION 1	POSITION 2	POSITION 3	POSITION 4	POSITION 5
DEPTH 1	4	4		4	1
DEPTH 2	3		5		
DEPTH 3	1	3	1		
DEPTH 4		1	4	3	5
DEPTH 5	2	2		3	4
TOTAL FOR EACH POSITION	10	10	10	10	10

MASTER KEY: 21244

Figure 9.5 Masterkeying by the numbers.

From position 2 on the chart, we see that four levers must be cut; all of these require a double gating cut. In position 3, four require a double gating and one needs widening. In position 5, only one requires another gating and four require widening at the current gate.

Another masterkeying method is to have what is known as a *master-tumbler lever* in each lever lock. The master tumbler has a small peg fixed to it that passes through a slot in the series tumblers. The master key raises the master tumbler. The peg, in turn, raises the individual change tumblers to the proper height so the bolt post passes through the gate in each lever. The lock is open.

This system should be ordered from the manufacturer. The complexities of building one yourself require superhuman skill and patience.

Masterkeying Disc Tumbler Locks

Disc tumbler locks have as few as three discs and as many as twelve. The most popular locks have five.

Figure 9.6 shows how the tumbler is modified for masterkeying. The left side of the tumbler is cut out for the master; the right side responds to the change key. The key used for the master is distinct from the change key in that its design configuration is reversed. The cuts are, of course, different. The keyway in the plug face is patterned to accept both keys. Since the individual disc tumblers are numbered from 1 to 5 according to their depths, it is easy to think of the master key and disc cuts on a 1 to 5 scale, but on different planes for both the key and the tumblers.

Uniform cuts are taboo. The series 11111 or 22222 would give very little security since a piece of wire could serve as the key. Other uniform cuts are out because they are susceptible to shimming. To keep the system secure, it is best to keep a two-depth interval between any two change keys. For example, 11134 is only one depth away from 11133, so 11134 should be used and 11133 omitted. The rationale is that 11134 is the more complex of the two.

The next step is to select a master key combination composed of odd numbers. At the top of your worksheet, mark the combination you have selected for the master key. Below it add a random list of possible change-key numbers. If you choose a systematic approach in developing change-key numbers, you compromise security. On the other hand, the systematic approach ensures a complete list of possible combinations. Begin systematically; then, randomly select the change-keys combinations.

A single code could be used for all customers. The main point to remember is to use different keyways.

Masterkeying Pin Tumbler Locks

Masterkeying is more involved than modifying the cylinder. It requires the addition of another pin sandwiched between a top and bottom pin. This pin, logically enough, is called the *master pin*.

The master system is limited only by the cuts allowed on a key, the number of pins, and the number of pin depths available. Since this book is for beginning and advanced students, the subject will be covered on two levels: the simple master key system for no more than 40 locks in a series and the more complex system involving more than 200 individual locks.

It is important to remember that a master key system should be designed in such a way as to prevent accidental cross keying.

Figure 9.6 The master key operates on the left side of the tumbler.

Pins are selected on the basis of their diameters and lengths. Master-pin lengths are built around the differences between the individual pin lengths. Consider, for a moment, the Yale five-pin cylinder with pin lengths ranging from 0 to 9 cuts. Each pin is 0.115 inch in diameter. Lower pin lengths are as follows:

0 = 0.184 inch	5 = 0.276 inch
1 = 0.203 inch	6 = 0.296 inch
2 = 0.221 inch	7 = 0.315 inch
3 = 0.240 inch	8 = 0.334 inch
4 = 0.258 inch	9 = 0.393 inch

As an illustration, let's masterkey ten locks with five tumblers each. Each tumbler can have any of ten different individual depths in the chamber.

Determine the lengths of each pin in each cylinder. Mark these down on your worksheet. The master key selected for this system may have one or more cuts identical to the change keys.

Cut a master key to the required depths. In this instance, load each cylinder plug by hand.

Using the known master key depth, subtract the depth of the change key from it. The difference is the length of the master key pin. If the change key is 46794 and the master key is 68495, the master pin combination will be as follows:

Chamber position	Bottom	Master
1	4	2
2	6	2
3	7	3
4	9	0
5	4	1

Follow this procedure for each plug. Notice that not all the chambers have a master pin. Such complexity is not necessary and makes the lock more vulnerable to picking. Each master pin represents another opportunity to align the pins with the shear line. Figure 9.7 is an extreme instance, with five master pins and three master keys.

In practice, locksmiths avoid most of this arithmetic by compensating as they go along. For example, an unmasterkeyed cylinder has double pin sets. Masterkeying means that an additional pin is added in (usually) position 1. This drives the bottom pin lower into the keyway. The bottom pin must be shortened to compensate.

A grand master key or a great-grand master key adds complexity to the system. You have two choices: add master pins in adjacent chambers (the Yale approach shown in Figure 9.8), or stack pins in the first chamber. Suppose you have a No. 4 bottom pin and the appropriate master pin (Figure 9.9A). In order to use grand and great-grand master keys, you must add two No. 2 pins so that all four pins will operate the lock (Figure 9.9B).

Master key inserted

Figure 9.7 Masterkeying pin tumbler locks means a combination other than the change key will raise the pins to the shear line.

CHANGE KEY INSERTED

GRAND MASTER INSERTED

MASTER KEY INSERTED

GREAT-GRAND MASTER INSERTED

Figure 9.8 The Yale great-grand master system requires five master pins.

Developing the Master Key System

A master key system should be planned to give the customer the best security that the hardware is capable of. Begin by asking the customer these questions:

- Do you want a straight master key system with one master to open all locks? Or, do you want a system with submasters? That is, do you want a system with several submasters of limited utility and a grand or great-grand master?

- What type of organizational structure is within the business? Who should have access to the various levels?

- How many locks will be in each submaster grouping?

- Is the system to be integrated into an existing system, or will the system be developed from scratch?

- What type of locks do you have?

Figure 9.9 Stacking pins is an alternative way of masterkeying.

Once you have considered these answers (with the help of your catalogs, specification sheets, technical bulletins, and experience), you are ready to begin developing a master key system.

The purpose of masterkeying is to control access. A key may open one lock, a series of locks, a group composed of two or more series, or every lock in the system.

Master key terms

Before we go much further into the subject, it is wise to spend a few minutes defining terms. This glossary was prepared by Eaton Corporation (Yale locks) and is reprinted with their permission.

Bicentric cylinder. A pin tumbler cylinder with two plugs, which effectively make it two cylinders in one. The bicentric cylinder is recommended for large, multilevel master key systems where maximum security and expansion are required.

Building master key. A master key that opens all or most of the locks in an entire building.

Change key (or individual key). A key that will usually operate only one lock in a series, as distinguished from a master key that will operate all locks in a series. Change keys are the lowest level in a master key system.

Changes (key). See *key changes*.

Construction breakout key (CBOK). A key used by the owner to make all construction master keys permanently inoperative.

Construction master key (CMK). A key normally used by the builder's personnel for a temporary period during construction. It operates all cylinders designated for its use. The key is permanently voided by the owner when he accepts the building or buildings from the contractor.

Control key. A key used to remove the central core from a removable core cylinder.

Controlled cross keying. See *cross keying.*

Cross keying. When two or more different change keys (usually in a master key system) intentionally operate the same lock.

Cross keying, controlled. When two or more change keys under the same master key operate one cylinder.

Cross keying, uncontrolled. When two or more change keys under *different* master keys operate one cylinder.

Department master key. A master key that gives access to all areas under the jurisdiction of a particular department in an organization, regardless of where these areas are in a building or group of buildings.

Display room key. A special hotel change key that allows access to only designated rooms, even if the lock is in the shut-out mode. With many types of hotel locks, this key will also act as a shut-out key, making all other change and master keys inoperative, except the appropriate individual display room key and the emergency key.

Dummy cylinder. One without an operating mechanism; used to improve the appearance of certain types of installations.

Emergency key (EMK). A special, usually top level, hotel master key that operates all the locks in the hotel at all times. An emergency key will open a guest room lock even if it is in the shut-out mode. With many types of hotel locks, this key acts as a shut-out key, making all other change and master keys inoperative, except the appropriate individual display room key and the emergency key.

Engineer's key (ENG). A selective master key used by various maintenance personnel to gain access through many doors under different master and grand master keys. The key can be set to operate any lock in a master key system and, typically, fits building entrances, corridors, and mechanical spaces. Establishing such a key avoids issuing high level master keys to maintenance personnel. See *selective master key.*

Floor master key. A master key that opens all or most of the locks on a particular floor of a building.

Grand master key (GM or GMK). A key that operates a large number of keyed-different or keyed-alike locks. Each lock is usually provided with its own change key. The locks are divided into two or more groups, each operated by a different master key. Each group can be operated by a master key only, or by the grand only, or by a master and the grand.

Great-grand master key (GGM or GGMK). A key that operates a large number of keyed-different and/or keyed-alike locks. Each lock is usually provided with its own change key. The locks are divided into two or more groups, each operated by a different grand master key. Each of these groups is further subdivided into two or more groups, each operated by a different master key. A group can be operated by the great-grand only, a grand only, a master only, or any combination of the three.

Guest room change key.   The hotel room key that is normally issued to open only the one room for which it was intended. The guest room key cannot be used to set a hotel lock in the shut-out mode from the outside of the room, nor will it open a hotel lock from the outside if the lock is in the shut-out mode.

Hotel-function shut-out.   When a hotel-function lock is in the shut-out mode, regular master keys of all levels and the guest room key will not open the lock from the outside of the room. Most hotel-function locks can be set in the shut-out mode with the thumbturn or pushbutton from the inside, or with the emergency or display room key from the outside.

Hotel keying.   Keying for hotel-function locks.

Hotel great-grand master keys, grand master keys, or master keys.   Depending on the level of the system, these keys function as they would in a normal system except they cannot be used to set a hotel lock in the shut-out mode from the outside of the room, nor can they open a hotel lock from the outside if the lock is in the shut-out mode.

Housekeeper's key.   A grand master key in a great-grand level hotel system that normally operates all the guest rooms and linen closets in the hotel.

Interchange.   See *key interchange*.

Key bitting.   A number that represents the depth of a cut on a tumbler-type key. A bitting is often expressed as a series of numbers or letters that designate the cuts on a key. The bittings on a key are the cuts which actually mate with the tumblers in the lock.

Key bitting depth.   The depth of a cut that is made in the blade of a key. See *root depth*.

Key bitting list.   A list originated and updated by the lock manufacturer for every master key system established. This list contains the key bittings of every master key and change key used in the system. Each time an addition is made to the system, all new bittings are added to the list. It is essential that a complete copy of this list be furnished to any personnel servicing a master key system locally. The lock manufacturer should be informed of any changes made locally to a keying system.

Key bitting or cut position (also called spacing).   The location of each cut along the length of a key blade. It is determined by the location of each tumbler in the lock. Bitting position is measured from a reference point to the center of each cut on the key. The most common reference point is the key stop, but the tip of the key is sometimes used.

Key change number.   A recorded number, usually stamped on the key for identification. A key change number can be either the direct bitting on the key or a code number.

Key changes (chges).   The total possible number of different keys available for a given type of tumbler mechanism. In master key work, the number of different change keys available in a given master key system.

Key interchange. An undesirable situation, usually in a master key system, whereby the change key for one lock unintentionally fits other locks in the system.

Key section (KS). The cross-sectional shape of a key blade that can restrict its insertion into the lock mechanism through the keyway. Each key section is assigned a designation or code by the manufacturer. A key section is usually shown as a cross section viewed from the bow towards the tip of the key. See *keyway*.

Key set (or set). A group of locks keyed exactly the same way. A key set is usually identified with a key symbol. See *standard key symbols*.

Keyed-alike (KA). A group of locks operated by the same change key. Not to be confused with masterkeying.

Keyed-different (KD). A group of locks each operated by a different change key.

Keying. A term, used in the hardware industry, that refers to the arrangement of locks and keys into groups to limit access.

Keying levels. The stratification of a master key system into hierarchies of access. Keying systems are available with one or more levels. The degree of complexity of the system depends on the number of levels used. Generally, the top level master key can open all locks in the system. Each successive intermediate level of master key can open fewer locks but can open more locks than a change key.

Keying system chart. A chart indicating the structure and expansion of a master key system, showing the key symbol and function of every master key of every level.

Keyway (Kwy). The shape of the hole in the lock mechanism that allows only a key with the proper key section to enter. See *key section*.

Levels. See *keying levels*.

Maid's key. A hotel master key, given to the maid, that gives access only to the guest rooms and linen closets in her designated area of responsibility. A hotel is normally divided into floors or sections with a different maid's key for each floor or section. A maid's key will not open a guest room if the lock is in the shut-out mode.

Multiple key section system (or sectional key sections). Used to expand a master key system by repeating the same or similar key bittings on different key sections. Keys of one section will not enter locks with a different section, yet there is a master key section milled so it will enter some or all of the different keyways in the system. See *simplex key section*.

Paracentric keyway. A keyway in a cylinder lock with one or more side wards on each side projecting beyond the vertical centerline of the keyway to hinder picking. See *simplex key section*.

Pin tumblers. Small sliding pins in a lock cylinder, working against drivers and springs and preventing the cylinder plug from rotating until raised to the exact height by the bitting of a key.

Plug (of a lock cylinder). The round part containing the keyway and rotated by the key to transmit motion to the bolt or other locking mechanism.

Plug retainer. The part of a lock cylinder that holds the plug in the shell.

Privacy key. A change key set up as part of a master key system but not operated by any master keys or grand masters of any level. This key is set up for such areas as liquor-storage rooms in hotels, narcotic cabinets in hospitals, and food storage closets where valuables are kept.

Removable core cylinder. A cylinder containing an easily removable assembly that holds the entire tumbler mechanism, including the plug, tumblers, and separate shell. The cores are removable and interchangeable with other types of locks of a given manufacturer by use of a special key called the control key. See *control key*.

Root depth. Refers to the distance from the bottom of a cut on a key down to the base or bottom of the key blade. Root depth is easy to determine since it measures the amount of blade remaining, rather than the amount that was cut away (bitting depth).

Selective master key. A special, top-level master key in a grand or great-grand system that can be set to operate any lock in the entire system, in addition to the regular floor or section master key, without cross keying. Typical selective master keys include an engineer's key (ENG), nurse's key (NUR), and attendant's key (ATT). The number of selective master keys is normally limited to one or two and should be set up when the original system is established. See *engineer's key*.

Simplex key section. A single independent key section that cannot be expanded into a multiple key section system. Simplex key sections, such as the Yale "Para," are used for stock locks and small master key systems.

Spacing. See *bitting position*.

Standard key symbols. A uniform way of designating all keys and cylinders in a master key system. The symbol automatically indicates the exact function of each key or cylinder in the system, without further explanation.

Tailpiece. The connecting link attached to the end of a rim cylinder that transmits the rotary motion of the key through the door into the locking mechanism.

Top level master key. The highest level master key in a multilevel keying system that fits most of the locks in the system.

Tumbler. One or more movable obstructions in a lock mechanism that dog or prevent the motion of the bolt or rotation or the plug and are aligned by the key to remove the obstruction during locking or unlocking.

Uncontrolled cross keying. See *cross keying*.

Visual key control. A system of stamping all keys and the plug face of all lock cylinders with standard key symbols for identification purposes. Other key and cylinder stamping arrangements are available but are not considered visual key control.

Standard key symbol code

Figure 9.10 illustrates keying levels of control and the rudiments of the standard key symbol code. Great-grand master keys are identified by the letters GGM. Grand mas-

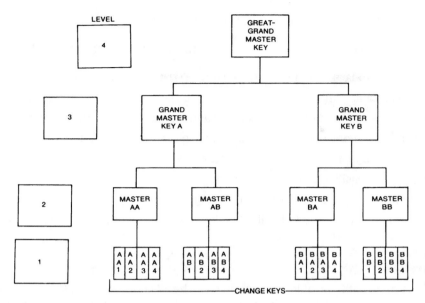

Figure 9.10 Key levels.

ters carry a single letter, beginning with the first in the alphabet and identifying the hierarchy of locks that the individual grand master keys open. Master keys carry two letters; the first identifies its grand master, the second identifies the series of locks under it. Thus, a master key labeled AA is in grand master series A and opens locks in master key series A. Master key AB is under the same grand master, but opens locks in series B. Master BA is under grand master B and opens locks in series A. Change keys are identified by their master key and carry numerical suffixes to show the particular lock that they open. Any key in the series can be traced up and down in the hierarchy. Thus, if you misplace change key AB4, you know that master key AB, grand master A, or the great-grand master will open the lock.

There are special keys that are out of series. Some of these keys are mentioned in the glossary, together with the appropriate key symbols for them.

If cross keying is introduced into the system—that is, if a key can open other locks on its level—the cylinder symbol should be prefixed with an X. If the cylinder has its own key, it is identified with the standard suffix. For example, XAA4 is a change-key cylinder that is fourth on this level. It may be cross-keyed with AA3 or any other cylinder or cylinders on this level. By the same token, master key AA, grand master A, and the great-grand master will open it. Elevator cylinders are often cross-keyed without having an individual change key. It is no advantage to have a key that will operate the elevator cylinder and no other in the system. These cylinders are identified as X1X, X2X, and so on.

The symbols that involve cross keying apply to the cylinder only; all other symbols apply to the cylinder and the key. This point may seem esoteric, but ignoring it causes the factory and everybody else grief. There is, for example, no such thing as an X1X key. Nor is there an XAA4 key. Change key AA4 fits cylinder XAA4 that happens to be cross-keyed with another cylinder. Key AA4 does not fit any cylinder except XAA4.

There are certain advantages to using the standard key symbol code:

- It is a standardized method for setting up the keying systems.
- It maintains continuity from one order to the next.
- It indicates the position of each key and each cylinder in the hierarchy.
- It helps to control cross keying, since each cross-keyed cylinder is clearly marked.
- It offers a method of projecting future keying requirements.
- It can be easily rendered on a chart.
- It allows better control of the individual keys within the system.
- It is a simple method of selling and explaining the keying system to the architect or building owner.

Selling the system

It is important to be able to communicate the advantages of the key symbol code and the implications it has for setting up an ordered, coherent, and secure keying system. This may take some selling on your part, since architects and building owners tend to think of keys and locks as individual entities and not part of a larger system.

Selling a comprehensive masterkeying system involves the following:

- Explaining the subject of masterkeying to the architect or owner
- Reviewing the plans of the building(s)
- Choosing the proper level of control required
- Selecting a keying system that will cover present and future expansion requirements
- Presenting the system to the architect or owner at a meeting dealing specifically with keying
- Recording all changes to the system as agreed at the meeting
- Marking plans with the proper key symbol on the side of each door where the key is to operate
- Presenting the owner with a schematic layout of the entire system and showing him the layout of the masters, grand masters, etc.

Master Key System Variations

So far, the structure of the master key system has been discussed. Within this broad structure, there are many opportunities for variations. Some of these variations involve the possible range of keyways; others involve special hardware such as removable cylinders, master-ring cylinders, and rotating tumblers. Each of these variations can extend the range, flexibility, and security of the system. A locksmith must be conversant with all of them.

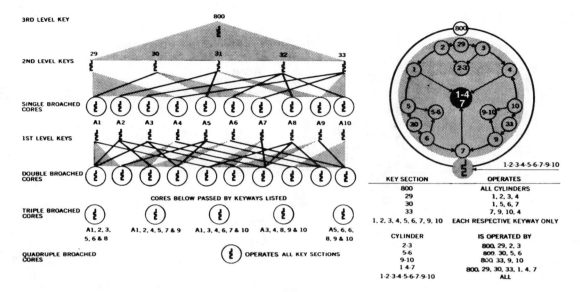

Figure 9.11 A keyway system developed by the General Lock Company.

Keyway variations

Figure 9.11 illustrates a system of control based on keyway design. General Lock's Series 800 key section passes all cylinders in the system; 29, 30, 31, 32, and 33 are submasters, each passing four cylinders; the change keys are restricted to their own individual cylinders. In its fullest expansion, the system includes 35 different keyways on the change-key level and four master key levels.

High-security pin tumbler cylinders

Figure 9.12 illustrates features of the General lock:

- 100° key bitting for long wear (A).
- Master pins have a minimum length of 0.040 inch (B). Shorter pins tend to wedge in the chamber.
- Only two pins are standard (C).
- Pins and springs are made of corrosion-resistant alloy (D and E).
- The keyway is part of the security system (F).
- If requested, the factory supplies special identification for keys, cores, and cylinders (G).

The Emhart (Corbin) High-Security Locking System uses rotating and interlocking pins (Figure 9.13). The pins must be raised to the shear line and, at the same time, rotated 20° so the coupling can disengage (Figures 9.14 and 9.15). Rotation is by virtue of the skew-cut bitting on the key (Figure 9.16). Figure 9.17 illustrates the way the cylinder is armored. The pins are protected by hardened rods and a crescent-shaped shield.

Figure 9.12 General locks have special features described in the text.

Figure 9.13 Corbin's High Security Locking System depends on split and rotating pins.

Figure 9.14 Pins are rotated in either direction to allow the joint to uncouple. (*Emhart Corp.*)

Figure 9.15 Pins must be rotated and brought to the shear line for the lock to open.

Figure 9.16 The Corbin key has a bitting cut at 20-degree angles.

PROTECTIVE
DRILL-RESISTANT
SHIELD

DRILL-RESISTANT
HARDENED STEEL
ARMOR RODS

CYLINDER PLUG HAS SIX
CIRCULAR GROOVES, CENTRAL
TO TUMBLER PIN HOLES

Figure 9.17 Passive defense measures include hardened steel pins and armor plate. (*Emhart Corp.*)

Removable-core cylinders

Removable-core cylinders are increasingly popular. Figure 9.18 illustrates the Corbin cylinder, a type typical of most. To rekey the change key, follow this procedure:

1. Obtain a Corbin rekeying kit. The kit includes the necessary pins, gauges, and tools.

2. Mount the cylinder in a vise.

TUMBLER SPRING
1. DRIVER PIN
2. CONTROL PIN
3. MASTER SPLIT PIN
4. BALL END TUMBLER PIN

RETAINER

CYLINDER CORE

FRONT PORTION OF
CORE REMOVED TO
SHOW INTERIOR DETAIL

LOCKING LUG PORTION
OF CONTROL SLEEVE

Figure 9.18   Corbin removable core cylinder.

3. Remove the plug retainer.
4. Select as the plug extractor the key with the deepest bitting. Normally, the grand master key meets this specification; however, there are instances where the engineer's key will have the deepest bitting. A shallow-cut key complicates matters by forcing the control pins, drivers, and buildup pins into the cylinder.
5. Withdraw the plug and remove all pins from their chambers. Figure 9.19 illustrates this procedure.
6. Determine the bitting of the change key.
7. Write down the new combination. As an example, suppose the original change-key bitting is 513525 and we wish to reverse it to 525315 (Figure 9.20).
8. Install the tumbler pins, ball end down.
9. Use a depth gauge to determine the master key bitting.
10. Calculate the master pins by subtracting the change-key bitting combination from the master key combination. If the master key combination was 525763, the difference between it and the new change-key combination would be 448.
11. Insert the master key into the plug.
12. Install the appropriate master pins (Figure 9.21).
13. Remove the master key carefully and insert the grand master key. Select the master split pins by subtracting the master key bitting from the grand master key bitting. All pins should be flush with the surface of the plug.
14. Assemble the plug and cylinder.
15. Test all keys.
16. Lubricate the keyway with a pinch of powdered graphite.

GRAND MASTER KEY

FOLLOWER

DUMP ALL PINS FROM THE PLUG

Figure 9.19 Using the appropriate follower, remove the plug, then dump the pins. (*Emhart Corp.*)

CHANGE KEY

5 2 5 3 1 5

Figure 9.20 Pin length corresponds to the change key combination. (*Emhart Corp.*)

MASTER KEY

4 4 8

Figure 9.21 Subtract the change key combination from the master key combination. The difference represents the length of the master pins. (*Emhart Corp.*)

Master-ring cylinders

The Corbin master-ring cylinder is shown in Figure 9.22. The change key operates the plug plungers, and the master key operates the plunger in the master ring. Sometimes called *two-in-one* cylinders, these cylinders increase the range of key combinations for any given system.

To rekey the change key, follow this procedure:

1. Mount the plug in a vise and remove the cylinder slide with a pair of pliers or a small chisel.

Figure 9.22 Corbin master ring cylinder.

2. Remove the springs, drivers, and pins (Figure 9.23).

3. Ream pin holes through the shell, master ring, and plug (Figure 9.24).

4. Assuming that the original combination was 414472, reversing the combination gives 274414. This will be the combination of the new change key.

5. Reverse the pins to conform with the new combination. That is, the pin that was first goes into the last chamber; the pin that was second goes in to the fifth chamber, and so on.

6. As you install each pin, tamp it home with a drill bit and turn the key. If the key will not turn, you have confused the pin sequence.

7. Assemble the lock and test.

Figure 9.23 After the slide is withdrawn, remove springs, pins, and top pins. (*Emhart Corp.*)

USE OLD CYLINDER AS A HANDLE

REAMER

Figure 9.24 Ream the pin holes through the shell, master ring, and plug. (*Emhart Corp.*)

To rekey the master key, follow this sequence:

1. Mount the cylinder in a vise and remove the cylinder slide with a pair of pliers or a small chisel.

2. Remove the springs, drivers, and pins.

3. Determine the master key bitting with a gauge. The combination runs from the shoulder to the tip of the key, the reverse of the usual sequence. As an example, let it be 678572.

4. Write down the change-key combination and reverse it. Suppose the combination is 275414. Reversed, it is 414572.

5. Subtract the reversed change-key combination from the master key combination (678572—414572). The difference is 264000.

6. Insert the master key into the cylinder and select the appropriate buildup pins. In this case the pins are 264. *Note:* If any number in the change-key combination is greater than the master key number above it, you must use a negative number buildup pin in the chamber. For example, a 678572 master key combination with an 814572 change-key combination requires a –2 buildup pin, together with a 6 and 4 pin.

7. As you install each buildup pin, seat it with a drill bit and turn the key to determine that you installed the correct pin.

8. Insert the drivers into the cylinder chambers. Either of two drivers are used, depending on the lock style. Spool drivers (No. J-172) are furnished with mortise cylinders; straight drivers (No. M-099) are used on other cylinders.

9. Insert the springs into their chambers.

10. Holding the springs down with your thumb, try the change key. Do the same for the master and grand master key.

11. Mount the cylinder in the slide, hammering the slide down for a secure fit. Be careful not to damage the threads on mortise cylinders.

12. Try all the keys.

A Simple Master Key System

When you are asked to install a large number of locks for a business or a large residence wherein the customer wants the locks masterkeyed, your expertise in the realm of the master key systems is put to the test.

Probably the hardest part of developing a master key system is creating a working system that takes into consideration any variables required by the customer. For this reason, many locksmiths do not perform their own masterkeying of systems; rather, they rely on the many years of experience, expertise, and professionalism from a factory-developed system. The use of a factory system ensures that the system meets all the criteria set forth by the customer, the cylinders are correct, and the possibility of extending the system in the future can be assured. For the locksmith, time, personnel, and money are saved through this process.

On the preceding pages, you have seen various aspects of masterkeying, the types of locks that can be used, the differences in the systems, and the potential for system expansion at a later date. I will not discuss the varied intricacies of such a system, but will view a small system that you may want to consider and keep in the back of your mind for small jobs that come up. These are jobs wherein an existing system may be masterkeyed, using the same cylinders, or a job where a residence is to be masterkeyed for a homeowner.

Figure 9.25 is a sample master key system developed for a three-floor office building. In addition to the individual office keys on each floor, you have the master key, extra cylinders for additional inner offices on each floor (or for use as replacement cylinders), a main door key, a key for the building maintenance shop, lavatory keys for both the men's and women's rooms on the three floors, and the start of an additional key code system you can develop should the building or other facility require more individual keys than the system was designed for.

Within this system, depending on the size of the installation you are developing, select the number of keys for individual locks to be used. In this case, we are talking about 20 individual door locks on each floor. The master key has already been designated. The progression in the development of the master key system is that the depths of the various individual keys are such that any given key can be re-cut to become a master key.

In actually rekeying the various cylinders, you will need the standard pins for each cylinder, and then a complete set of master pins. In this case, though, you may choose to use the lower ranking pins of the current system. Why? Because the cylinder pins are worn in addition to the interior of the cylinder itself. Also, in this case, it will save you time and money in having to order the pins.

What you are doing is determining for each key the depths that will be required for each pinning. As an example, let's take two cylinders from the first floor for rekeying. The first cylinder, 522534 will have the following pinning:

Regular Pin: 4 - 2 - 1 - 4 - 3 - 2
Master Pin: 1 - 1 - 1 - 1 - 4 - 2

With these pins in the cylinder, with the driver pins and tension spring included, either the individual or the master key will open the lock.

For the second cylinder in the series, 522534, the pinning is as follows:

Regular Pin: 4 - 2 - 1 - 4 - 4 - 2
Master Pin: 1 - 1 - 1 - 1 - 3 - 2

Again, with these pins, both the individual and the master key will operate the lock.

Now, how were these particular pins determined? The best way to figure this out is to lay out a simple chart, as shown in Figure 9.26. As you progress in selecting pins for the individual cylinders, first mark the appropriate points on the chart that indicate the master key number. (You can preprint this on a form and save a lot of time.) Then, as you select each cylinder to work with, put down on the chart the code number of that particular key (Figure 9.27). The difference between the two points, if

Master: 431472

First floor	Second floor	Third floor
522534	652534	752534
522544	652544	752544
522554	652554	792554
522564	652564	792564
522574	652574	792574
522594	652584	792584
522504	652594	792594
522524	652504	792504
522634	652624	792624
522644	652634	792634
522654	652644	792644
522664	652654	792654
522674	652664	792664
522684	652674	792674
522694	652684	792684
522604	652694	792694
522624	652734	792604
522724	652744	792734
522734	652754	792744
522744	652764	792754

Extra cylinders for each floor:

552784	652514	752614
522514	652524	752601
522614	604714	792602

Main door key: 346572
Bld maintenance shop key: 434672
Construction key: 466662

Note: If building were to have four or more floors, then the last three
      digits for the various individual key could be as above, but the first three digits would be:

Fourth floor: 514
Fifth floor: 614
Sixth floor: 714

If each floor is to have two lavatories, they could be keyed as follows:

First floor:	731573	331577
Second floor:	831573	331578
Third floor:	931573	331579

Figure 9.25 Sample master key system for a three-floor building.

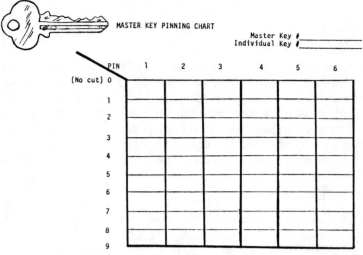

Figure 9.26 Sample chart used for arranging individual cylinder pinning for a master key system.

Figure 9.27 Layout chart with master key and individual key cylinder pinning indicated.

any, is the difference between the code number variations. This means that the pinning difference must be the same.

In Figure 9.27, for column six, notice that the individual key requires a 4 pin, but the master key requires a 2 pin. The difference of 2 is obvious. The maximum allowable height for the overall height of both pins combined must be 4, so you require a 2 pin for the master key to operate this particular pin, but also a 2 pin for the individual key. If you only had the 2 pin in the lock, only the master key would be able to open the cylinder at this point; by adding the second 2 pin, you make it possible for the individual key to also open the lock.

If you continue onward, you can readily see how each individual pinning is to be accomplished, what pins will be required, and the insertion sequence into the lock cylinder.

10

Detention Locks and Hardware

The detention industry includes private, county, state, and federal sectors. It provides many opportunities for locksmiths who understand the special needs of the industry. This chapter describes the various types of detention facilities and shows how to select and install hardware for each type.

To recommend the right locks, electric strikes, and other hardware, you have to consider the security level of the facility as well as that of specific areas within the facility. Security levels include low, medium, high, and maximum. It's important to understand that a low-security facility, for example, can have low- and high-security areas. But its high-security areas might not have the same needs as a high-security area in a maximum security facility.

Types of Locks

Mechanical and electromechanical locks are commonly used for high-security applications. Typically, the mechanical models have wafer or lever tumblers and are mortise-mounted in a 1¾-inch, hollow metal door with steel bolts and shear-proof inserts. Electromechanical locks are installed in frame jambs and can be operated remotely from an electric console. They require 24 volts of direct current (Vdc) and 110 volts of alternating current (Vac).

The mogul cylinder is often used for medium-security applications (Figure 10.1). It has stainless steel pin tumblers and can be masterkeyed. Most moguls offer four keying levels: Day Key/Pass Keys, Master Key, Grand Master Key, and Great-Grand Master Key.

Standard security hardware and mortised pin tumbler locks are used for low-security applications. The locks are typically installed in 2-inch-wide jambs and come in solenoid and motor versions. (Because of its superior sideload capabilities, the motor version is specified for medium-security applications.)

Figure 10.1 The Maxi-Mogul is a popular high-security cylinder for detention facilities. (*Folger Adam Company*)

Electric Strikes

Heavy-duty electric strikes are often used in detention centers because the devices provide a means of both mechanical and electric access control. A low- or medium-security detention facility might use them for individual confinement rooms, but they're not a good choice for individual cells in a maximum security prison.

Electric strikes are often used to create a *mantrap system*—a way of controlling access to a restricted area by using two electrically interlocked doors. A mantrap requires a person to walk through and close one door before the second door can be opened. That allows a prisoner to be inspected, searched, or detained before gaining access to a more restricted area. Highly sophisticated mantraps incorporate closed-circuit televisions (CCTV), intercoms, and security guards.

A mantrap uses a door monitor switch at each door with a connecting switch to the electric strike power leads of the opposing door. The switch acts as a power inhibitor to the opposing strike, so when one door is opened power is cut to the second door. Only after the first door has been closed can the second be opened. While the second door is open, the first door remains locked.

Choosing the Right Equipment

In addition to knowing the security levels of the facility and its specific areas, you need to answer the following questions before choosing equipment:

- What types of doors need to be secured? (sliding, swinging, wood, metal, etc.)
- What functions must the locks perform?
- Do the locks need to operate mechanically or electromechanically?
- Do the locks need to be motor or solenoid operated?

The answers to these questions will help you choose among equipment such as that described in the following pages.

Southern Steel 802 electric deadbolt. The 802 is a mortise lock that provides swinging doors with auxiliary security locking and unlocking from a remote location (Figure 10.2). It can be unlocked electrically from remote control or by key switch or push-button at the door. It automatically deadlocks when the door is closed, but it has no protruding deadlatch mechanism accessible to inmates when the door is open.

The lock is nonhanded and is activated by dual 115 Vac solenoids for power locking and unlocking. The case and cover are made of cold-rolled steel plate. It measures 4⁹⁄₁₆ inches × 1¾ inches × 9 inches and weighs 9 pounds. The lock's stainless steel bolt is ⅝ inch in diameter with a ⁹⁄₁₆-inch throw.

Special versions include:

- 802KR, which is equipped with a mogul cylinder for mechanical unlocking by key at the door and activated by dual 115 Vac solenoids for power locking and unlocking
- 802FL, which is activated by one 24 Vdc solenoid for power unlocking and has a bolt that is extended automatically when power is interrupted (fail secure)
- 802FU, which is activated by one 24 Vdc solenoid for power locking and has a bolt that retracts automatically when power is interrupted (fail secure)

Southern Steel 1010 deadlock. The 1010 lock is a key-operated lock designed for swinging doors on access panels, plumbing chases, control cabinets, and other areas that require only minimum security. Figure 10.3 shows a model 1010 deadlock. According to Southern Steel's technical manuals, the lock shouldn't be used on exterior doors that directly contact the inmate population. Some jails do ignore that advice.

The lock has four holes in its corners that are used for mounting the lock to a door and securing the cover plate on the lock case. A fifth screw holds the cover plate in place; that screw hole is hidden behind a sticker on the case. You have to peel up the lower right corner of the sticker to gain access to the screw hole.

Figure 10.2 The 802 Electric deadbolt provides swinging doors with auxiliary security locking from a remote location. (*Southern Steel*)

Figure 10.3 The model 1010 deadlock is for swinging doors on access panels, plumbing chassis, control cabinets, etc. (*Southern Steel*)

The lock measures 4¼ inches × 1¼ inches × 3 inches and weighs 4½ pounds. The cover is made of cold-rolled steel plate and the case is malleable iron. The bolt is yellow brass and measures 1½ inches × ¾ inch with a ⅝-inch throw.

Southern Steel 1050. The 1050 is for swinging and sliding exterior gates. Standard mounting is for chain link fencing but can be modified for other types of gates. The lock can work by a remote pushbutton switch that activates a solenoid that raises an internal bolt, enabling the gate to open. It remains unlocked until the gate is closed, then automatically locks. The bolt can also be retracted manually with a key.

The lock cover and case are made of cold-rolled steel plate; its latchbolt is cold-rolled steel. The lock measures 7 inches × 3¾ inches × 11 inches and weighs 38 pounds. Its five tumblers are 0.097-inch spring temper brass actuated by phosphor bronze springs.

Figure 10.4 The 10195B Electrolock is for electric or manual control of swinging doors in low security inmate housing areas. (*Southern Steel*)

Southern Steel 1080 deadlock. The 1080 is for swinging cell doors, corridor doors, storage room doors, and high-security doors where slam-locking isn't required. The lock is key operated and deadlocks in both the locked and unlocked position. It measures 5½ inches × 1½ inches × 3¾ inches and weighs 7 pounds. The cover is made of cold-rolled steel plate; the case is malleable iron. The lock's bolt is cold-rolled steel with ¼-inch diameter, case-hardened, steel insert pins; it measures 2 inches × ¾ inch and has a 1⅛-inch throw.

Southern Steel 10125 institutional mortise lock. The 10125 is for swinging doors in low-security housing areas where electric or manual operation is desired. A door position switch, door closer, and power transfer hinge are recommended. The lock can work by using a remote pushbutton switch to activate a solenoid that locks and unlocks knobs. The lock bolt can also be retracted mechanically by mogul cylinder and knobs. The knobs can be locked out and unlocked by the mogul cylinder. When the knobs are locked out, the lock automatically deadlocks on closing.

The lock measures 4⅝ inches × 1¼ inches × 9½ inches and weighs 11 pounds. Its cover, case, face plate, and mogul cylinder are made of yellow brass. Its bolt is made of stainless steel and measures 1⅜ inches × ⅝ inch with a ⅝-inch throw.

Southern Steel 10195B electrolock. The 10195B is for electric and manual control of swinging doors in low-security inmate housing areas (Figure 10.4). It's jamb mounted and utilizes an optional builders hardware cylinder. A door position indicator switch, door closer, and heavy-duty door pull are recommended accessories.

The lock measures 1½ inches × 1²³⁄₃₂ inches × 13⅜ inches and weighs 6½ pounds. Its bolt is made of nickel aluminum bronze. It measures ⅝ inch diameter and has a ⅝-inch throw. The lock case is made of die-cast brass and the face plate is made of brass.

In its electrical function (24 Vdc, 2½ amp), the lock can be unlocked from a remote console or by a key switch or pushbutton at the door. It automatically deadlocks when the door is closed. In its mechanical function the bolt is extended and retracted by key at the door (just like any other key-operated mechanical lock). When key internal switch is used, however, the key will only retract the bolt.

Standard features of the lock include:

- No protruding deadlatch mechanism accessible to inmates when door is open
- Designed for mortise jamb mounting in standard door frame without visible lock pocket
- Automatic deadlocking of bolt
- Activated by power-surge dual solenoids
- During power outage, bolt remains retracted mechanically without relying on continuous electric power to its solenoid
- Bolt position indicator switch
- Nonhanded (except when internal key switch is specified)

Special features include

- Key switch on either side or both sides
- Fail unlock version available

- Master key override—one key operates internal switch for electric bolt retraction and the other key retracts bolt mechanically
- Cylinder extenders for locks keyed stop side

Institutional hinges

202 strap hinge, 2 inches. Use 202 hinges for shutter and wicket doors where a light duty security hinge is required. They're furnished with blank mild steel leaves for welded installation. A 202 hinge measures 4⅛ inches × 2 inches × $\frac{5}{32}$ inch and weighs 0.5 pounds. Its pin is cold-rolled, case-hardened steel and measures $\frac{5}{16}$-inch diameter.

203 special hinges, 3 inches. The 203 is for food passes, wicket doors, observation shutters, and other small swinging doors where a medium-duty hinge is needed. It's available with solid leaves for welded application or predrilled for fasteners. The 203FP is a version of the 203FS with a built-in stop to hold the food pass door in a horizontal position for use as a shelf.

The 203 measures 2¾ inches × 3 inches × ¼ inch and weighs 0.9 pounds. The hinge leaves are made of mild steel. The hinge pin is cold-rolled, case-hardened steel and measures $\frac{5}{16}$-inch diameter.

204 institutional hinges, 4½ inches. Available in full mortise and half mortise configurations, 204 hinges can be used on swinging hollow metal or wood doors (Figure 10.5). The hinge leaves are made of die-cast brass to provide maximum strength and durability. Each hinge features a ⅝-inch diameter stainless steel, nonremovable pin, plus two sets of hardened steel ball bearings and races. Each unit weighs 1.6 pounds and comes with ¼-20 brass, flat-head safety screws.

Figure 10.5 The 204 Institutional Hinges have a nonremovable hinge pin. (*Southern Steel*)

204E electric power transfer hinge. Designed to supply power from door frames to electric locks on hollow metal doors, the 204E power transfer hinge contains five completely concealed and tamper-resistant, Teflon-coated conductors. It's available only in full mortise configuration to be used with 204FM institutional hinges. The unit has a 1 amp capacity, 40 volts maximum. It measures 4½ inches × 4½ inches × $\frac{3}{16}$ inch and weighs 1.6 pounds. Its hinge leaves are made of die-cast brass. The hinge pin is stainless steel and has a ⅝-inch diameter.

Buying and Selling Safes

Virtually every consumer and business has documents, keepsakes, collections, or other valuables that need protection from fire or theft. But most people don't know how to choose a device that meets their protection needs, and they won't get much help from salespeople at department stores or home improvement centers. By knowing the strengths and weaknesses of various types of safes, you will have a competitive edge over such stores.

No one is in a better position than a knowledgeable locksmith to make money selling safes. Little initial stock is needed, they require little floor space, and safes allow for healthy price markups. This chapter provides the information you need to begin selling safes to businesses and homeowners.

Types of Safes

There are two basic types of safe: fire (or *record*) and burglary (or *money*). *Fire* safes are designed primarily to safeguard their contents from fire, and *burglary* safes are designed primarily to safeguard their contents from burglary (Figure 11.1). Few low-cost models offer strong protection against both potential hazards. That's because the type of construction that makes a safe fire-resistant—thin metal walls with insulating material sandwiched in between—makes a safe vulnerable to forcible attacks. The construction that offers strong resistance to attacks—thick steel walls—causes the safe's interior to heat up quickly during a fire.

Most fire/burglary safes are basically two safes combined, usually a burglary safe inside a fire safe. Such safes can be very expensive. If a customer needs a lot of fire and burglary protection, you might suggest that he or she just buy two safes. To decide which type of safe to recommend, you need to know what your customer plans to store in it.

Figure 11.1 A UL-listed fire safe or chest is constructed so that its interior stays below a certain temperature during a fire. Nonrated containers might provide no more fire protection than a metal box offers. (*Sentry Group*)

Safe Styles

Fire and burglary safes come in three basic styles, based on where the safe is designed to be installed. The styles are wall, floor, and in-floor. *Wall* safes are easy to install in homes and provide convenient storage space (Figure 11.2). Such safes generally provide little burglary protection when installed in a drywall cutout in a home. Regardless of how strong the safe is, a burglar can simply yank it from the wall and carry it out. To provide good security, a wall safe needs not only a thick steel door, but to be installed with concrete in a concrete or block wall.

Floor safes are designed to sit on top of a floor. Burglary models should either be over 750 pounds or bolted in place. Figure 11.3 shows a popular floor safe. One way to secure a floor safe is to place it in a corner and bolt it to two walls and to the floor. (If you sell a large wall safe, make sure your customer knows that the wheels should be removed from the safe.)

Figure 11.2 A wall safe can be a convenient place to store documents. (*Gardall Safe Corporation*)

Figure 11.3 A floor safe can provide more usable storage space if it has movable shelves. (*Gardall Safe Corporation*)

Figure 11.4 An in-floor safe can provide strong resistance to burglary and fire when properly installed in a concrete basement floor. (*Sentry Group*)

In-floor safes are installed below the surface of a floor (Figure 11.4). Although they don't meet construction guidelines to earn a UL fire rating, properly installed in-floor safes offer a lot of protection against fire and burglary. Because fire rises, a safe below a basement floor won't quickly get hot inside. For maximum burglary protection, the safe should be installed in a concrete basement floor, preferably near a corner. That placement makes it uncomfortable for a burglar to attack the safe.

Make sure your customer knows that he or she should tell as few people as possible about the safe. The fewer people who know about a safe, the more security the safe provides.

Special Safe Features

Important features of some fire and burglary safes include relocking devices, hard-plate, and locks. Relocking devices and hardplate are useful for a fire safe but are critical for a burglary safe. If a burglar attacks the safe and breaks one lock, the *relocking devices* automatically move into place to hold the safe door closed. *Hardplate* is a reinforcing material strategically located to hinder attempts to drill the safe open. Never recommend a burglary safe that doesn't have relockers and hardplate.

Safe locks come in three styles: key-operated, combination dial, and electronic. *Combination dial* models are the most common (Figure 11.5). They are rotated clockwise and counterclockwise to specific positions. *Electronic* locks are easy to operate and provide quick access to the safe's contents. Such locks run on batteries that must be recharged occasionally. For most residential and small business purposes, the choice of a safe lock is basically a matter of personal preference.

Underwriters Laboratories Fire Safe Ratings

UL fire safe ratings include 350-1, 350-2, and 350-4. A 350-1 rating means the temperature inside the safe shouldn't exceed 350°F during the first hour of a typical home fire. A safe rated 350-2 should provide such protection for up to two hours. Safes with a 350-class rating are good for storing paper documents because paper chars at 405°F. Retail prices for fire safes range from about $100 to over $4000. Most models sold in department stores and home improvement centers sell for under $300.

Underwriters Laboratories Burglary Safe Standard

The UL 689 standard is for burglary resistant safes. Classifications under the standard, from lowest to highest, include Deposit Safe, TL-15, TRTL-15x6, TL-30, TRTL-30, TRTL-30x6, TRTL-60, and TXTL-60. Figure 11.6 shows a depository safe. The classifications are easy to remember when you understand what the sets of letters and numbers mean. The two-set letters in a classification (TL, TR, and TX) signify the type of attack tests a safe model must pass. The first two numbers after a hyphen represent the minimum amount of time the model must be able to withstand the attack. An additional letter and number (e.g., X6) tells how many sides of the safe have to be tested.

The TL in a classification means a safe must offer protection against entry by common mechanical and electrical tools, such as chisels, punches, wrenches, screw-

Figure 11.5 Combination dial locks are popular for safes. (*Sargent & Greenleaf, Inc. Reprinted with permission*)

Figure 11.6 A depository safe has a slot in it that allows a store clerk to drop money in it. (*Gardall Safe Corporation*)

drivers, pliers, hammers and sledges (up to 8-pound size), and pry bars and ripping tools (not to exceed 5 feet in length). TR means the safe must also protect against cutting torches. TX means the safe is designed to protect against cutting torches and high explosives.

For a safe model to earn a TL-15 classification, for example, a sample safe must withstand an attack by a safe expert using common mechanical and electrical tools for at least 15 minutes. A TRTL-60 safe must stand up to an attack by an expert using common mechanical and electrical tools and cutting torches for at least 60 minutes. A TXTL-60 safe must stand up to an attack with common mechanical and electrical tools, cutting torches, and high explosives for at least 60 minutes.

In addition to passing an attack test, a safe must meet specific construction criteria before earning a UL burglary safe classification. To be classified as a deposit safe, for example, the safe must have a slot or otherwise provide a means for depositing envelopes and bags containing currency, checks, coins, and the like into the body of the safe, and it must provide protection against common mechanical and electrical tools.

The TL-15, TL-15, and TRTL-30 safe must either weigh at least 750 pounds or be equipped with anchors and instructions for anchoring the safe in a larger safe, in concrete blocks, or to the premises in which the safe is located. The metal in the body must be the equivalent to solid open-hearth steel at least 1 inch thick having an ultimate tensile strength of 50,000 pounds per square inch (psi). The TRTL-15x6, TRTL-30x6, and TRTL-60 must weigh at least 750 pounds and the clearance between the door and jamb must not exceed 0.006 inch. A TXTL-60 safe must weigh at least 1000 pounds.

TL-15 and TL-30 ratings are the most popular for business uses. Depending on the value of the contents, however, a higher rating may be more appropriate. Price is the reason few companies buy higher rated safes. The retail price of a TL-30 can exceed $3500. A TLTX-60 can retail for over $18,000.

Such prices cause most homeowners and many small businesses to choose safes that don't have a UL burglary rating. When recommending a nonrated safe, consider the safe's construction, materials, and thickness of door and walls. Better safes are made of steel and composite structures (such as concrete mixed with stones and steel). Safe walls should be at least ½ inch thick and the door at least 1 inch thick. Make sure the safe's boltwork and locking mechanisms provide strong resistance to drills.

Buying Safes

Safes vary greatly in quality, utility, and price. Even with your special knowledge, you need to buy high-quality safes at the lowest prices possible because you're also competing against other locksmiths. You need to decide whether to buy directly from a manufacturer or a distributor. The advantage of buying from a manufacturer is a lower price when ordering volume. Most locksmith shops don't purchase enough safes to benefit from the potential savings. The advantages of dealing with a distributor are the option of buying in small quantities and (usually) faster delivery.

Although it isn't necessary to stock a lot of safes, you'll need to have a few models on display if you want to be taken seriously as a safe seller. Most people want to see a product before buying it. Initially you shouldn't stock large, heavy safes because they're expensive, hard to transport, and usually don't sell quickly.

12

Keyed Padlocks

Padlocks are used for securing outbuildings, bicycles, buildings under construction, tool boxes, paint lockers, and even automobile hoods. Because of this wide use, a locksmith needs a good working knowledge of padlocks.

While padlock exteriors vary, the functional and operational differences are few and are similar to those of other locks. Padlocks might use pin tumblers, wards, wafers, levers, or a spring bar. Some must be shackled closed before the key can be removed; this feature is made possible by a spring-loaded coupling. Key security is improved and there is less likelihood of leaving the lock open.

Choosing a Padlock

Ask the customer if he or she has a brand preference, then ask the following questions:

- Are width, case length, and shackle clearance crucial (Figure 12.1)?
- Where will the lock be used?
- How often will the lock be opened? The price of the lock has a direct relationship to its wearing qualities.
- Is the lock intended to secure valuable property? An inexpensive lever lock is adequate to keep children from straying into the backyard but inappropriate for a boat trailer.
- Will the lock be used indoors or outdoors? If it is an outdoor lock, will it be protected from the elements?

Figure 12.1 Padlock nomenclature.

Warded Padlocks

Warded padlocks locks have limited life spans, particularly when used outdoors. The cheapest locks of this type are only good for a few thousand openings and, when locked, give minimal security.

Most of these locks have three wards, although the cheaper ones have only two. Figure 12.2 illustrates the principle. The key must negotiate the wards before it can disengage the spring bar from the slot in the shackle end. Keys are flat or corrugated. The latter is a mark of the Master Lock Company.

Repairs to the lock itself are out of the question, because it would be cheaper to buy a new one.

Pass keys

Figure 12.3 shows how a pass key is cut to defeat the wards. The broad tip of the key opens the spring bar; locks with two spring bars require a key cut as shown in Figure 12.4.

Figure 12.2 A warded padlock. Section A represents a rotating disc keyway; section B one of the three wards; and section C the spring bar.

Figure 12.3 Wards and their limitations. View A shows the ward arrangement and the necessary key bitting; view B illustrates a pass key.

DOUBLE SPRING BAR

Figure 12.4 A pass key for a lock with two spring bars.

Some corrugated keys can be reversed to fit other locks by filing on the back of the blades. Alternately, you can file all the unnecessary metal off, converting the key into a pass key.

In many localities, pass keys are illegal and, unless you are a locksmithing student from an accredited school, a locksmith trainee, or a licensed locksmith, possession of such a key is a criminal offense. If you have the need for a pass key, keep it in a safe place in your home or office.

Key cutting

A warded padlock key is simple to duplicate. Follow this procedure:

1. Using the original key as a guide, select the appropriate key blank.

2. Smoke the original key and mount the original and the blank in your vise.

3. Using a 4-inch warding file, cut away the excess metal until the blank is an exact copy of the original.

4. Turn the keys over and repeat the operation.

5. Remove the burrs from the duplicate and test it in the lock. If it sticks, the cuts are not deep or wide enough. Make the appropriate alterations and you will have a perfect duplicate key.

Key impressioning

Impressioning a warded key should take less than five minutes. Follow this procedure:

1. Select the appropriate blank and thoroughly smoke it.
2. Insert the key and twist it against the wards. Do this several times to get a clear impression.
3. Mount the key in the vise.
4. Make shallow cuts where indicated.
5. Smoke the key again and try it in the lock.
6. Remove the key and file the cuts as indicated.
7. Continue to smoke, test, and file until the key turns without protest. Do not go overboard with the file. If the bits are too deep, the key will work but may break off in the lock.

Wafer Disc Padlocks

Wafer disc padlocks are recognized by their double-bitted keys and can be a headache to service. If you do not have a key, the lock can be opened by either of two methods:

- Picking is possible but it takes practice. Purchase a set of lock picks for these locks and spend a few hours learning the skill.
- You can try the lock with a set of test keys, available from locksmith supply houses.

Disassembly

If the lock is already open or a key is available, your job is almost half done. Release the retaining ring clip with a length of stiff wire inserted into the toe shackle hole. Remove the cylinder plug and lay out the parts.

Keys

Duplicate keys can be cut by hand or with a machine designed for this purpose. These machines are expensive. Locksmiths get around the problem by stocking cylinder inserts and precut keys. The insert replaces the original wafer mechanism. A precut key requires that the discs be realigned to fit. You will need some blank discs and should have access to the cutting tool described in the chapter on double-bitted locks and keys.

In extreme cases, you can insert the key and file the disc ends for shell clearance. This is not the sort of thing that a real professional would do, but it works.

Do not attempt to cut double-bitted keys by the impressioning method. A new plug and key is the better choice.

Pin Tumbler Padlocks

There are hundreds of pin tumbler locks on the market, but most fall into two categories, based on the case construction: laminated or extruded.

Laminated padlocks

Slaymaker pin tumbler padlocks are made up of a series of steel plates held together by four rivets at the casing corners. Follow these service procedures:

1. Use a hollow mill drill to shear off the rivet heads on the case bottom; this technique allows reuse of the rivets.
2. Remove the bottom plate.
3. Remove the entire cylinder section in one piece.
4. Two cylinder-housing types are used. The most popular requires a follower tool to keep the pins intact.
5. Make the necessary repairs.
6. Insert the cylinder into the casing and replace the bottom plate over the four rivet ends.
7. After checking the action, use a ball-peen hammer and repeen the rivet ends.

Extruded padlocks

Extruded locks are made from a single piece of metal, usually brass. You will need the following tools to service these locks:

- Small nail with the point cut off
- Key blanks
- Pliers
- Small punch
- Hammer
- Small light (overhead lights are too bright; a flashlight is acceptable)

To disassemble the extruded lock, first locate the plug pins or cover on the bottom of the case. Since the case is highly polished, this is sometimes hard to do. Tilting the lock under the light will show a faint outline of the cover cap or plug pins.

To assemble extruded locks held together by exposed pins, follow this procedure:

1. Determine the diameter of the plug pins.
2. Select a drill bit smaller than the diameter of the pin.
3. Drill slowly and only a fraction of an inch deep—the pin does not extend very far into the lock. If you drill through it, you can damage the spring.
4. Extract the plug.

5. Remove each spring and pin.

6. Remove the cylinder. The cylinder is held by a pin that may be covered by a small pin plug. Rapping the lock against a hard surface may shock the plug loose. If not, dislodge the pin plug with a small screwdriver.

7. Make the necessary repairs.

8. Assemble the lock.

9. Fill the pin holes with brass wire.

10. File the wire flush with the case and polish.

Extruded locks with a retaining plate over the pins require a slightly different procedure. If the lock is open, drive the plate out with a punch inserted at the shackle hole. If it is locked, follow this procedure:

1. Drill a small hole in the plate off-center.

2. Using an ice pick or other sharp instrument, drive the plate down toward the shackle. This will buckle the plate and cause the edges to rise.

3. Pry the plate loose by working a sharp instrument around the edges.

4. If the plate is not severely damaged, it can be reused. Bow the plate slightly so that the center bulges outward when the plate is installed.

5. Peen the edges of the plate to form a tight seam.

Other locks mount the plug and cylinder by means of horizontal pins running across the width of the lock. These pins can be seen under strong light.

Key fitting for pin tumbler padlocks

Follow this procedure:

1. Insert the blank into the plug.

2. Place the No. 1 pin in its chamber.

3. Using a blunted nail as a punch, drive the pin into the blank. A single light hammer tap is enough to impression the blank.

4. Remove the blank.

5. File the blank at the impression.

6. Insert the key and try to turn the cylinder. File as necessary.

7. Move to the next pin and repeat the process.

Major Padlock Manufacturers and Their Products

American (Junkunc Brothers)

All American (Junkunc Brothers) padlocks share the same patented locking device—two hardened steel balls fitting into grooves in the shackles. This arrangement is the best ever devised. Applying force on the shackle wedges the balls tighter. The H10 model, for example, requires more than 5000 pounds to force; test locks have been stressed to 6000 pounds and still worked. All H10 locks use a ten-blade tumbler. (The exceptions to these statements are those locks with a deadlock feature. The key must be turned to lock the shackle. These padlocks are made on an entirely different principle and are not discussed here.)

American padlocks have a removable cylinder that simplifies servicing. If a customer wants keyed-alike locks, the modification takes only a few minutes. Locksmiths can also key-alike different models.

These locks have three basic subassemblies: cylinder assembly, locking mechanism assembly, and the shackle assembly.

Cylinder removal and installation.   Follow this procedure:

1. Open the lock, exposing the *retaining screw* at the base of the *shackle hole*.
2. Remove the retaining screw (I in Figure 12.5). If the screw is stubborn, use penetrating oil and let it set a few moments before attempting to turn the retaining screw.
3. Strike the side of the lock with a leather mallet. The purpose is to force the *retaining pin* (C) into the space vacated by the retaining screw. Referring to the drawing, note how the pin retains the cylinder in the slot shown.
4. Withdraw the key together with the cylinder. Place a new cylinder in the case.
5. Assemble the lock in the reverse order of disassembly. *Note:* The H10 series uses a plated silver cap dimensionally identical to the brass caps on the other models.

A. Key
B. Cylinder
C. Cylinder-Retaining Pin
D. Retainer Washer
E. Retainer Assembly With Coiled Spring
F. Hardened Steel Balls
G. Case
H. Shackle Spring
I. Retainer Screw
J. Shackle or "Staple"

Figure 12.5 A popular American padlock.

The locking mechanism. The brass *retainer* is the heart of the assembly (E in Figure 12.5). Retainers vary in size according to the lock model. A30 and AC20 retainers have a groove milled in the body. L50, K60, and KC40 retainers are plain. The H10 is much larger than the others.

The retainer assembly has three parts: a washer, coil spring, and retainer body. The *washer* (D) fits over the retainer. One end is hooked to accept the coil spring. The other end of the spring is moored in a hole on the retainer.

To remove the retainer, follow this procedure:

1. Grasp the retainer body (E) with needle-nosed pliers.
2. Rotate the retainer 45° and pull.
3. Remove the steel balls.

To replace, follow these steps:

1. Grease the balls so they will stay put.
2. Replace the balls and spread them apart.
3. Exert pressure against the shackle to hold the balls.
4. Determine that the washer is correctly aligned with the spring. The free end of the spring has to be in line with the retaining-pin hole in the case.
5. Place the retainer assembly about halfway in the case with the washer riding in the grooves provided. The end of the spring also rides in the groove.
6. Twist the retainer about a quarter turn to the right and down.

Shackle. The various models have different shackle lengths and diameters. The spring (H in Figure 12.5) must match. For example, an L-shaped spring is used on the L50 lock. An extra pin guides the long spring on the K60. The shackle in the H10 is secured with a hardened steel pin; other models secure the shackle with the retainer mechanism.

Pin tumbler padlocks. The American five-pin tumbler padlock was designed so the cylinders can be changed quickly (Figure 12.6). Late production 100- and 200-series use the same cylinder assembly and keyway, reducing the inventory load.

Assembly and disassembly of the pin tumbler padlock is only slightly different from that of the other American padlocks. To change the cylinder, follow this procedure:

1. Open the padlock to expose the retaining screw at the base of the shackle.
2. Remove the retaining screw with a small screwdriver.
3. Pull the cylinder out of the case. *Note:* Leave the lock unlocked; do not depress the shackle.
4. Insert a new cylinder in the case. Replace the screws, bringing the cylinder almost flush with the case.

A. Key
B. Cylinder Assembly
C. Retainer Washer
D. Retainer Assembly
E. Hardened Steel Balls

F. Case or Body
G. 6/32 Brass Screw
H. Shackle Spring
I. Shackle or "Staple"

Figure 12.6 An American pin tumbler padlock.

To assemble the locking mechanism:

1. Assemble the retainer (D in Figure 12.6) and the retainer washer (C) with the free end of the spring tightly against the left side projection on the washer.

2. Place the shackle spring (H) in the hole on the end of the shackle (I).

3. Insert the shackle spring into the deep well of the lock body.

4. Drop the balls (E) into the case bottom and move them into the pockets with a small screwdriver. There must be room for the retainer assembly.

5. Depress the shackle so the balls cannot slide back into the center of the case.

6. Insert the retainer assembly (C and D). The retainer washer (C) fits into the elongated hole in the case.

7. Make sure the assembly is bottomed in the hole.

8. Insert a retainer-assembly tool into the cylinder hole so the step on the tool engages the retainer step. Turn the tool clockwise until the lock opens.

9. Install the cylinder assembly. Holding it flush with the case bottom, insert the 6 × 32 brass screw (G) into the shackle hole and tighten it snugly.

Note: The American Lock Company will provide the retainer-assembly tool free for the asking.

Because of the close tolerances of American padlocks, they will not operate properly in extreme cold weather unless Kerns ML3849 lubricant is used. Graphite is acceptable in milder climates.

To remove the cylinder in the 600 series pin tumbler padlocks, unlock the padlock and turn the shackle as shown in Figure 12.7. Depress the spring-loaded plunger (A) with a small screwdriver until the plunger is flush with the wall of the shackle hole. At the same time, pull on the key. Withdraw the cylinder.

B SHACKLE

C SPRING-LOADED PLUNGER

KEY A

D

CYLINDER

Figure 12.7 The American 600 series padlock.

Master padlocks

The Master Lock Company makes a variety of locks, ranging from simple warded locks to sophisticated pin tumbler types.

The Super Security Padlock. An improved version of the familiar Master padlock was recently introduced—the Master Super Security Padlock (Figure 12.8). It was designed to give greater security than standard locks and is recommended for warehouses, storage depots, and industrial plants, as well as the home and yard. Figure 12.9 is a cutaway view of the mechanism. Salient features include:

- A patented dual-lever system to secure the shackle legs. Each lever works independently of the other and is made of hardened steel.
- The long shackle leg is tapered to align the locking levers.
- For weather protection, the case laminations are made of hardened steel and are chrome-plated.
- For better protection, the case is larger than that of standard locks.
- A rubberoid bumper prevents the lock body from scratching adjacent surfaces. This feature is not found on other locks.
- Tension tests show that this high-security lock can tolerate without damage a force on the shackle of 6000 pounds.

When servicing the Super Security lock, follow this procedure:

1. Drill out the bottom rivets.
2. Remove the plates, one at a time, and keep them in order.
3. Remove the lock plug.
4. Assemble the lock using new rivets.

Figure 12.8 The Master Super Security Padlock.

Figure 12.9 A cutaway of the Master Super Security Padlock.

Master Lock's Pro Series locks. Master Lock Company's Pro Series is a complete line of high-security padlocks that are especially for commercial use. The line includes "Weather Tough," "High-Security Shrouded," and a round, solid steel lock (Figures 12.10 through 12.12). All have black bodies and shackles made from a hardened boron steel that provides up to 15,000 pounds of cut resistance. Inside, the shackles are locked with dual ball-bearing locking mechanisms that resist up to 13,000 pounds of pulling and prying. In addition, they are shrouded in solid iron.

The cylinders of all Pro Series locks feature special spool pins to make them pick-resistant. All Pro Series padlocks are rekeyable with cylinder and shackle options that are quick and easy to change.

Figure 12.10 Pro Series "Weather Tough" padlocks are protected with "Xenoy" thermoplastic covers and feature boron alloy shackles and special drain channels that move water and debris through the lock body and away from the locking mechanism. (*Master Lock Company*)

Figure 12.11 Pro Series High Security shrouded padlocks have a steel shackle guard that protects the shackle from saws, bolt cutters, and prying. (*Master Lock Company*)

Figure 12.12 One type of Pro Series High Security padlock comes in a round, solid steel nonshrouded version. (*Master Lock Company*)

The Pro Series also features a line of Weather Tough locks especially for outdoor applications. These locks have the same boron alloy shackles and dual ball-bearing locking mechanisms found on the High-Security models. A xenoy thermoplastic cover protects against key and cylinder jamming from sand, dirt, and other contaminants. Special drain channels inside move water and debris through the lock body and away from the locking mechanism for trouble-free opening.

Removing and replacing the laminated and Pro Series cylinder

1. Unlock the padlock and remove the key (Figure 12.13). Turn the shackle away from its opening. *Note:* If rekeying WO (without cylinder) padlocks, use a screwdriver to open the shackle. Insert the tip of the screwdriver through the opening in the bottom plate. Turn the mechanism as shown to trip the shackle release. (For Pro Series 6000 locks simply pull up on the shackle. Cylinder extension and ball bearings are included in a separate polybag.)

Figure 12.13 Exploded view of a laminated Master padlock. (*Master Lock Company*)

2. The cylinder retainer screw is located inside the shackle opening. Using an appropriately sized hex wrench, rotate the screw counterclockwise (left) until the cylinder retainer door separates from the padlock body.

3. Remove the retainer door and nut. Slide the cylinder out. Be sure to cradle the padlock in your hand to catch parts.

4. Place the new cylinder rear-end first into the padlock body. Replace the retainer door and nut.

5. Tighten the retainer screw with a hex wrench.

Removing and replacing the cylinder in solid-body padlock (Nos. 220, 230)

1. Follow Steps 1 and 2 of "Removing and Replacing the Laminated and Pro Series Cylinder."

2. Slide the cylinder retainer out (Figure 12.14). Turn the retainer upside down to remove the cylinder.

3. Place the new cylinder with the rear end facing up into the retainer. Slide the retainer into padlock body.

4. Tighten the retainer screw with a hex wrench.

Rekeying the Pro Series cylinder

1. Examine the rear of the cylinder and note the flat indent on the otherwise round rim of the cylinder shell. This is the *index mark*. While holding the cylinder, insert the key and rotate it 90° toward you (counterclockwise) so the bottom of the keyway lines up with the index mark. *Caution:* Don't rotate the keyway 180° or the pins may jam.

Shackle

Hex wrench

Retainer
screw

Case

(Rear)

(Front)

Cylinder

Cylinder retainer

Key

Figure 12.14 Exploded view of a
solid-body Master padlock.
(*Master Lock Company*)

2. Line up the cylinder assembly tool shoulder to shoulder with the rear end of cylinder.

3. Carefully push the tool into the cylinder until the plug comes out. Leave the tool inserted in the cylinder shell to prevent top pins and springs from falling out.

4. Remove the key and turn the plug upside down so that bottom pins drop out.

5. Using a key cut gauge, determine the bottom pin sizes to match the new key cut.

6. Insert pins with pointed side down into the plug.

7. Insert the new key. Each pin should be flush with the plug surface.

8. Slowly push the plug back into the cylinder shell to force the assembly tool out the opposite end.

9. Carefully rotate the key 90° away from you (clockwise) back to the starting position. *Caution:* Don't rotate the keyway 180° or pins may jam.

Removing and replacing the Pro Series shackle

1. Follow steps 1 and 2 of "Removing and Replacing the Laminated and Pro Series Cylinder."

2. Remove the retainer door and nut. Slide the cylinder out.

3. Close the shackle.

4. Strike the bottom of the lock against the palm of your hand. The shackle extension and ball bearings will drop out.

5. Pull up on the shackle to remove it.

6. Insert the new shackle into the lock body.

7. To replace the ball bearings, turn the padlock upside down. Drop one ball bearing into the opening. Add a small amount of grease to lubricate if necessary. Use a hex wrench to carefully push the ball into the left channel.

8. Repeat with the other ball bearing, pushing it into the right channel.

9. Replace the cylinder extension with the large steel end in first, lining it up with the opening (make sure the ball bearings stay in either channel). Use a hex wrench to push the extension in as far as possible.

10. Place the cylinder—rear end first—into the padlock body (remove the key from the cylinder). Replace the cylinder retainer door and nut.

11. Use the key to open the lock and pull up on the shackle.

12. Insert the retainer screw into the shackle opening and tighten with a hex wrench.

ILCO

ILCO secures the cylinder with what the trade calls the "loose rivet" method. A single brass rivet extends through the case and into the cylinder. The rivet is headless (hence the term loose) and is secure by virtue of its near invisibility (Figure 12.15).

To remove the pin, follow these steps:

1. Hold the case up to the light and look for the shadow created by the pin. The pin is located ⅛ inch from the bottom of the case and is centered on the side.

2. Using a No. 48 bit, drill a hole in the end of the pin.

3. Thread a small screw into the hole.

4. Gently pry up on the screw and remove it together with the pin.

To insert a new pin, follow these steps:

1. Select a piece of brass from stock that exactly matches the diameter of the pin hole.

2. Use the next size smaller rod (to make removal easier if it is ever necessary) and cut it so that ½₂ inch stands out.

3. Peen over the end of the rod.

4. Burnish the case to camouflage the pin.

LOOSE-RIVET PADLOCK

Figure 12.15 ILCO loose-rivet padlock.

Another ILCO variant is a *cylinder retaining plate* on the bottom of the case. The rivet that holds the assembly together is cunningly disguised in the maker's name stamped on the plate.

To remove the retaining plate, follow this procedure:

1. Draw a line under the ILCO name.

2. At a right angle to the first line, draw a second line (on the outside edge of the O in ILCO).

3. Drill a shallow hole to accept the tip of a 6 × 32 machine screw. Fit the screw with a washer and thread it into the hole.

4. Using the screwhead for purchase, pry the retainer out of the case.

5. Withdraw the cylinder.

6. When the cylinder has been serviced and replaced into the case, mount the retainer in its original position, tapping it home with a flat-ended punch. Fill the hole with brass stock.

Kaba security padlocks

The Kaba KP 2008 security padlock (Figure 12.16) is a good quality, heavy-duty lock with an interior dust cover which, when coupled with the stainless steel body, means it can be used in any weather conditions. The dimensions are in Figure 12.17.

The KP 9 cable padlock has a plastic-coated steel cable. It is well suited to securing bulky items such as equipment on building sites, bicycles, etc. (Figure 12.18).

The KP 10 padlock, which has a shackle load of over two tons (Figure 12.19), is designed for strength. Its dimensions are shown in Figure 12.20.

The Epilok security hasp and staple (Figure 12.21) is a unique unit. A strong hasp and staple of hardened steel construction uses the unique Epilok spigot lock with its advanced Mini-Kaba locking cylinder. This unit will stand up to the most strenuous attack.

Figure 12.22 breaks down the unit a little more. The rawbolt staple (on the right) is separate from hasp unit and locking spigot lock. After the hole is drilled and the rawbolt inserted, the staple is screwed on tight, making removal exceptionally difficult. The security of this system is much greater than that of the several screws one would expect with a security hasp. The dimensions are provided in Figure 12.23.

The 45201 security window/patio door lock has a universal locking mechanism that can be used for any application where a sliding door or window requires additional locking on the top and/or bottom part of the frame. It is also very suitable for hinged windows or pivoting windows (Figure 12.24).

The lock is operated by a Kaba-20 cylinder mechanism. The same key can be linked to an unlimited number of locks with the same key code, including rim locks, padlocks, or mortise locks. All windows, patio doors, front and back doors can be quickly locked, every time. There is no need to search for individual keys on a large key ring. The dimensions for the window/patio door lock are in Figure 12.25.

These last two locks are excellent counter displays for the locksmith because they practically sell themselves to the customers. There is no need to provide any form of sales pitch for the product. They are top-of-the-line, with proven quality, reliability, and durability, and will give your customer years of valuable protective service.

Figure 12.16 Heavy-duty padlock with the Kaba 20 cylinder. (*Kaba Locks, Ltd.*)

Figure 12.17 Specification data for the heavy-duty padlock. Notice the two different shackle sizes available. (*Kaba Locks, Ltd.*)

Figure 12.18 Cable padlock suitable for securing small pieces of equipment and bicycles. (*Kaba Locks, Ltd.*)

Figure 12.19 The KP 10 has a large shackle into a solid locking body for increased strength. (*Kaba Locks, Ltd.*)

Figure 12.20 KP specification data. (*Kaba Locks, Ltd.*)

Figure 12.21 The Epilock high-security hasp and staple unit used the Mini Kaba cylinder. (*Kaba Locks, Ltd.*)

Figure 12.22 Epilok unit broken down into separate sections for installation. (*Kaba Locks, Ltd.*)

Hasp

Rawbolt staple

Staple

Figure 12.23 Specification data for the Epilock. (*Kaba Locks, Ltd.*)

Figure 12.24 Windows or patio door security is assured with this lock unit. (*Kaba Locks, Ltd.*)

Figure 12.25 Specification data for the window/patio lock. (*Kaba Locks, Ltd.*)

Helpful Hints

- It's easier to work with padlocks that are open. If the lock is not open, pick it open. Sometimes the shackle bolt can be disengaged with the help of a hatpin inserted through the keyway. However, locks are getting better and this technique doesn't work as well as it used to.

- Polish the lock before returning it to the customer. Use emery paper to remove the deep scratches, then burnish with a wire wheel. Finish by buffing.

- The best security in a padlock comes when the shackle is locked at both the heel and toe. The *double bolt action* (or *balls*) is the ultimate in padlock security, making it nearly impossible to force the shackle.

- When picking fails (and even the best of locksmiths may occasionally have this problem), use penetrating oil. Yale, Corbin, and ILCO pin tumbler locks are especially susceptible to penetrating oil.

13

Home and Business Services

Over the years a locksmith acquires a great deal of knowledge, much of which is learned from correcting his or her own mistakes. This is a hard school. The best way to solve problems is to never let them arise in the first place.

This chapter includes some basic techniques. Some of this material might seem obvious and you might wonder why it's in here. The answer is that these techniques have been tested and approved by the experts, and they can save you time, money, and the embarrassment of callbacks.

The following few pages are supplied courtesy of Corbin (Emhart Corporation). The first section applies to all locksets, regardless of make. Other sections detail service and troubleshooting procedures for specific Corbin locksets as well as other makes of locks.

Common Problems and Troubleshooting

If you encounter difficulties in the operation of a lockset, first review this checklist of common problems and solutions to see if you can clear up the difficulty.

- Is the door locked (Figure 13.1)?

- Are you using the right key (Figure 13.2)?

- If the latchbolt or deadbolt does not engage or disengage the strike or binds in the strike, it is usually due to bolt-strike misalignment.

- Has the door warped (Figure 13.3)?

- Is the door binding? Frames that are out of plumb are frequently the cause of faulty operation of locksets and binding of bolts in the strike (Figure 13.4).

- Are the hinges loose? Tighten the screws, filling holes if necessary, or rehang the door if the screws will not hold (Figure 13.5).

- Are the hinges worn? If excessive wear has occurred on the hinge knuckles, the door will not be held tightly. Replace the hinges.
- Is the frame sagging? If sag cannot be corrected and the door and frame returned to plumb relationship, planing or shaving the door and repositioning or shimming the strike may relieve this condition.
- When a key operates the latchbolt or deadbolt with difficulty, it is usually due to bolt-strike misalignment (Figure 13.6).

Figure 13.1 Is the door locked?

Figure 13.2 Are you using the right key?

Figure 13.3 Has the door warped?

Figure 13.4 Is the door binding?

Figure 13.5 Are the hinges loose?

Figure 13.6 File to correct bolt-strike misalignment.

Corbin Cylindrical Locksets

Common problems with Corbin cylindrical locksets include the following:

Problem 1: Latchbolt will not deadlock. Caused by deadlocking latch going into strike. Either the strike is out of line or the gap between the door and jamb is too great. Re-align the strike or shim the strike out towards the flat area of the latchbolt (Figure 13.7).

Problem 2: Latchbolt cannot be retracted or extended properly. Caused by latchbolt tail and latchbolt retractor not being properly positioned (Figure 13.8). Remove the lockset from the door. Reinsert the latchbolt in the door. Looking through the hole in the door, the tail should be centered between the top and bottom of the hole. Remove the latchbolt and insert the lockcase. Looking through the latchbolt hole in the

Figure 13.7 Latchbolt will not deadlock.

Figure 13.8 Latchbolt cannot be retracted or extended properly.

lock face of the door, the latchbolt retractor should be centered in the hole. Adjust the outside rose for proper position. Rebore the holes, if necessary, to line up the retractor and tail.

Problem 3: Latchbolt will not project from the lock face. Latchbolt tail and retractor may be misaligned (Figure 13.9). See Problem 2. If this is not the cause, the spring is probably broken.

Problem 4: Key works with difficulty. Lubricate the keyway (Figure 13.10). Do *not* use petroleum products. Spray powdered graphite into the cylinder or place powdered graphite or lead pencil shavings on the key. Move the key slowly back and forth in the keyway. Bitting (notches) on the key may be worn.

Corbin heavy-duty cylindrical locksets: service procedures

Servicing these locks is not difficult if the task is approached methodically.

Figure 13.9 Latchbolt won't project from lock face.

Figure 13.10 Lubricating keyway.

Figure 13.11 Tightening the lockset.

Tightening the locksets.   Tighten the inside rose thimble with a wrench. If the thimble needs to be taken up a great deal, tighten the *outside* rose at the same time to prevent possible misalignment and binding of both (Figure 13.11).

If the lockset is still not tight, back off the thimble. Using a screwdriver, push down the knob retainer and remove the knob and rose. If the spurs on the back of the rose are bent, straighten and reposition the rose so the spurs are embedded in the door. If the spurs are broken off, insert a rubber band or nonmetallic washer under the rose. Tighten the thimble.

Removing and installing the locksets

1. Remove the key from the knob.
2. Loosen the inside rose thimble with a thimble wrench. Pull the lock slightly to release the rose spurs from the door.
3. Disengage the inside knob retainer with a thimble wrench and pull out the inside knob.
4. Slide the rest of the lockset from the outside of the door.
5. Remove the bolt face plate screws and slide the latch unit from the lockface of the door.

To reinstall the lockset, reverse the procedures above. When inserting the case and keyed knob from the outside of the door, be sure the bolt retractor in its case properly engages the latchbolt tail.

Removing and replacing the cylinder in the locksets

1. Follow the procedure in the "Removing and Installing the Locksets" section.
2. Remove one case screw and slightly loosen the other.
3. Swing out the outside knob retainer (it pivots on the slightly loosened case screw).
4. Using a screwdriver, pry the knob filler cover off the outside knob.
5. Using a special Waldes ring pliers No. 3, remove the large Waldes ring from the groove and withdraw the shank and cylinder (Figure 13.12).

Figure 13.12 Removing cylinder.

To reinstall the cylinder, reverse the procedure. Be sure the knurled side of the Waldes ring is face up.

Changing the hand of the locksets. Remember that the lock cylinder should always be in position to receive the key with the bitting (notches) facing upward.

1. Follow the procedure in the preceding section.
2. Slightly loosen *one* case screw and back off the other (Figure 13.13).
3. Swing out the outside knob retainer (it pivots on the slightly loosened case screw).
4. Lift the knob out and rotate it 180°. Replace it in the case. Swing the knob retainer back into place.
5. Insert the case screw and tighten both screws.
6. Reinstall (Figure 13.14).

Figure 13.13 Changing lockset hand.

Figure 13.14 Reinstalling knob.

Figure 13.15 Unsnapping rose from rose liner.

Figure 13.16 Tightening rose.

Corbin standard-duty cylindrical locksets: service procedures

Tightening the locksets

1. Depress the knob retainer and remove the inside knob.
2. Unsnap the rose from the rose liner (Figure 13.15).
3. Place the wrench into the slot in the rose liner and rotate it clockwise until it's tight on the door. If the lockset is extremely loose, tighten the outside rose and inside liner *equally* (Figure 13.16).
4. Replace the inside rose; depress the knob retainer and slide the knob onto the spindle until the retainer engages the hole in the knob shank.

Removing locksets

1. Depress the knob retainer and remove the inside knob.
2. Unsnap the rose from the rose liner.
3. Place the wrench into the slot in the rose liner and rotate counterclockwise until it's disengaged from the spindle.
4. Remove the lock by pulling the outside knob.
5. Remove the bolt after removing the screws in the lockface of the door.

Reinstalling locksets

1. Adjust the lock for the door thickness by turning the outside rose until the edge of the rose matches one of the lines marked on the shank of the knob. (The first line is for a 1⅜-inch door, the second for a 1¾-inch door.) When using trim rosettes, increase the adjustment to compensate for the thickness of the metal.

2. Install the latchbolt in the face of the door.

3. Install the lock from the outside of the door so the case engages the slot in the latchbolt and the tail interlocks the retractor.

4. Replace the rose liner, rose, and knob on the inside of the door.

Reversing the knobs

1. Lock the exterior knob (Figure 13.17).

2. Hold the latchbolt in the retracted position with the key (Figure 13.18). Depress the knob retainer with a screwdriver through the slot in the knob shank.

3. Pull the knob off the spindle (Figure 13.19). Rotate the knob so the bitting on the key will be *up*.

4. Replace the knob by lining up the lance in the neck of the knob with the slot in the spindle.

5. Push the knob on the spindle until it hits the retainer button. Depress the retainer button and push the knob until it snaps into position.

Figure 13.17 Locking exterior knob.

Figure 13.18 Retract latchbolt.

Figure 13.19 Pull knob off spindle.

Removing the cylinders

1. Remove the outside knob. Turn the key in either direction until it can be partially extended from the plug (Figure 13.20).
2. Hold the knob and turn the key to the left, pulling slightly on the key until the cylinder disengages.

Corbin unit locksets (300 and 900 series)

Troubleshooting

Problem 1: Latchbolt binds or rattles in strike.   Adjust the strike. Because the lockset is preassembled as one unit, there are no internal adjustments to make. Check that the cutout for the lockset is square and at the right depth so the face of the lockset is flush with the face of the door. Then, adjust the nylon adjusting screw in the strike (Figure 13.21).

Problem 2: Key does not activate the knob or latchbolt.   Check for a worn key. If bitting (notches) in the key is worn down or the key is bent, the locking mechanism will not operate properly. If the key is not worn, spray powdered graphite into the keyway or put graphite or lead pencil filings on the key and move it back and forth slowly in the keyway (Figure 13.22). Never use petroleum products. Check that a binding latch is not the cause.

Problem 3: Lockset is loose in the door.   Tighten the *escutcheon* screws (Figure 13.23). Be sure to tighten them evenly.

Problem 4: Lockset has the wrong bevel for the door.   Reverse the lockset. Unit locksets with horizontal keyways may be changed from right- to left-hand regular bevel, or vice-versa by merely turning the lock upside down (Figure 13.24).

Figure 13.20 Remove outside knob.

Figure 13.21 Adjusting screw in strike.

Figure 13.22 Lubricating keyway.

Figure 13.23 Tightening escutcheon screws.

Figure 13.24 Reversing lockset.

Service procedures. Usually the unit locksets are easier to service than the cylindrical locksets. To service this series of locksets, proceed as follows:

Removing the locksets from a door. First, remove the through-bolts on the inside of the door (Figure 13.25). Push the outside escutcheon away from the door so the lugs clear the holes. Slide the assembly out of the door.

Removing 900 series knobs

1. Remove the attaching screws (Figure 13.26).
2. Snap off the dust cover.
3. Pry the wire retaining ring from the knob retaining key located in a slot in the frame tube. Remove the retaining key. Remove the inside knob, which is fastened to the knob shank.
4. Remove the inside escutcheon.
5. Loosen the escutcheon on the outside of the lock by inserting a screwdriver through the access hole from the inside of the lock frame; remove the screw and escutcheon fastener.
6. Pry the wire retaining ring from the retaining key located in a slot in the frame tube. Remove the retaining key. Remove the outside knob, which is fastened to the knob shank.

Figure 13.25 Removing through-bolts.

Figure 13.26 Removing attaching screws.

Removing cylinders

1. Using a screwdriver, pry the knob filler cover off the outside knob.
2. Use No. 103F92 Waldes retaining pliers to remove the Waldes ring.
3. Remove the shank and lock cylinder. Rekey the lock.

   Reverse the procedure to reinstall. Be sure that the beveled edge of Waldes ring faces away from the knob, and that the ring is properly seated in the groove (Figure 13.27).

Removing 300 series knobs

1. Remove the key from the lock (lock should be unlocked). Remove the three attaching screws (Figure 13.26).
2. Snap off the dust cover.
3. On the inside knob side, pry the wire retaining ring away from the knob retaining key located in the slot in the frame tube. Remove the retaining key.
4. Remove the knob that has the shank assembled to the knob.
5. Remove the inside escutcheon.
6. Loosen the escutcheon on the outside of the lock by inserting a screwdriver through the access hole from the inside of the lock chassis; remove the screw and washer.
7. Repeat step 3 on the outside knob and shank.

Corbin mortise locksets (7000, 7500, 8500 series)

Troubleshooting

Problem 1: With door open, latchbolt doesn't extend or retract freely. Check for binding against the rose. Adjust the knob. Loosen the roses or trim on the door. If the bolt now operates freely, the roses or trim must be realigned. A *knob aligning tool* is recommended. Check installation templates for proper position (Figure 13.28). If the bolt does not operate properly with trim and roses loosened, remove the lockset from the door. If the lockset operates properly when removed from the door, use a chisel to make the mortise larger so the lockset enters freely.

Problem 2: With door closed, latchbolt doesn't extend or retract freely, or door won't latch at all. Open the door. If the latchbolt still doesn't operate properly, see Problem 1.

Problem 3: Latchbolt "stubs" on the strike lip. Bend the strike slightly back toward the jamb. Wax or paraffin makes an excellent lubricant, as does silicone spray (Figure 13.29).

Problem 4: Deadbolt doesn't enter the strike. This is probably due to misalignment of strike and bolt, particularly in cases where door sag has taken place. Both latchbolt and deadbolt holes in the strike must be filed or the strike repositioned. Do *not* force the thumbpiece if the deadbolt doesn't extend and retract in the strike freely (Figure 13.30).

Figure 13.27 Removing cylinder.

Figure 13.28 Checking latchbolt.

Figure 13.29 Bending strike.

Figure 13.30 Filing strike.

   Problem 5: Removing the cylinder. Loosen the cylinder locking screws in the face of the lockset. If the scalp covers the set screws, remove the scalp. Unscrew the cylinder. When replacing, be sure the locking screws are firmly seated (Figure 13.31).

   Problem 6: Key does not operate the latchbolt or deadbolt. Loosen the cylinder locking screws. The cylinder is in the wrong position in the door so the cylinder cam does not engage the locking mechanism properly. Turn the cylinder a whole turn to the left or right until it works properly. The keyway must always be in position to receive the key with bitting (notches) up. Tighten the cylinder locking screws.

Figure 13.31 Removing cylinder.

Problem 7: Key turns hard when retracting the deadbolt. If the bolt operates freely with the door open, check the bolt-strike alignment. Check the scalp to be sure it is not binding the bolt or that paint over the bolt is not causing the bind.

Problem 8: Key works hard in the cylinder. Lubricate the cylinder with powdered graphite or place graphite on the key and move it back and forth slowly in the keyway. Never use any petroleum lubricants (Figure 13.32).

Problem 9: Key breaks in the lock. Remove the cylinder from the lock. Insert a long pin or wire into the back end (cam end of the cylinder). Move it back and forth until the broken key stub is forced out through the front of the cylinder. Clean the cylinder with ethyl acetate and lubricate it with graphite before reinstalling (Figure 13.33).

Problem 10: Thumbpiece trim doesn't retract latchbolt completely or doesn't extend bolt completely. Check for binding at inside trim; or, if the outside thumbpiece trim is used in conjunction with the inside panic device, check to see if it is operating properly and is not dogged down. Remove the thumbpiece and check the position of thumbpiece in relation to the latch trip at the bottom of the mortise lock case. When properly installed, the top of the thumbpiece should be up against the bottom of the latch trip but not lifting it (Figure 13.34). If the bolt doesn't retract fully when the thumbpiece is pushed down, the thumbpiece is too low on the door. Move the trim up as needed. If the bolt doesn't extend completely when the thumbpiece is released, the thumbpiece is too high on the door. Move the trim down as needed. If the trim is fixed and cannot be moved, carefully bend the thumbpiece tail up or down.

Service procedures. Mortise locksets are not difficult to work with. Servicing them involves only a few basic procedures.

Figure 13.32 Lubricating keyway.

Figure 13.33 Lubricating cylinder.

Figure 13.34 Installing thumbpiece.

Installing and adjusting the working trim.   Unscrew the spindle. Remove the mounting plates, roses, and thimble from the spindle. Unscrew the sleeve from the inside knob. Remove the spindle from the knob (Figure 13.35).

Install the rose-attaching plates. On wood doors, install flange washers through the mortise, turning the attaching plates into them. It may be necessary to mortise for the washer to permit the mortise lock to clear it. Install the mortise lock.

Align the plates using a No. 028 aligning tool. If this tool is not available, assemble the spindles allowing a ⅟₁₆-inch gap between halves of the swivel spindles with mounting plates, sleeve spacer, and adjusting nut in position as shown in Figure 13.35. Tighten the adjusting nut with your fingers. Mark screw holes through both mounting plates. Install the screws.

Figure 13.35 Removing spindle from knob.

Reassemble the spindle, tightening the adjusting nut, and back off one quarter-turn to line up with the flats of the spacer. The spindle should have a slight end chuck. Try the knob. If the latch binds, back off the adjusting nut another quarter-turn. Disassemble the spindle.

Assemble the roses and thimble. Reassemble the spindle complete with knobs. Tighten the rose thimbles and inside adjusting plate with furnished spanner wrenches. Knobs should turn freely in either direction from either side.

Installing and adjusting the lever handle trim assembly

1. Check that the handle hole is 1⅛ inch wide. On wood doors, remove the mortise lockset and make additional mortise for retainer on the inside of the case mortise. Replace the locksets and secure with screws.
2. Place the rose assembly spindle into the lock hub. The label shows the top and latch edge. Spot four screw holes through the assembly plate (Figure 13.36). Make sure that the latchbolt operates freely. Drill holes very carefully.
3. Attach the rose assembly to the door. (With a wood door, attach through the door to the retainer.)
4. Put on the cover plates. Place the lever handle on the assembly and secure with an Allen set screw.
5. Adjust. If the latchbolt binds, remove the lower plate and slightly change the rose assembly position.

Reversing the hand. As you will remember from Chapter 7, the hand refers to the position of the hinge on the door and to the direction of swing. Some locksets are universal and fit all four hands. Others must be modified in the field.

Figure 13.36 Installing lever handle.

1. Refer to Figure 13.37. Unfasten the two cap screws (A) and remove the cap.

2. Reverse the latchbolt (B). If the lock has an antifriction latch, first remove the L-shaped pin (C) and reinstall it on the opposite side after reversing the bolt. Hold the short leg against the side of the case with the pressure-sensitive decal.

3. Reverse the auxiliary latch (D) if an auxiliary latch is furnished. To do this, first remove the auxiliary latch lever and spring, then invert the latch so the concave surface is toward the stop in the frame. Reassemble the lever and spring.

4. Reverse the front bevel. Loosen three screws (E) in the top and bottom edges of the case, adjust to the proper bevel (or to flat), and retighten the screws securely.

5. Reverse the knob hubs (F): hub with ⅜-inch square spindle hole toward the outside.

6. Replace the cap and securely tighten the screws.

7. Check for proper operation before installing in the door.

Figure 13.37 Reversing the hand.

The Medeco Ultra 700 Deadbolt

Instructions for mounting the Medeco Ultra 700 Deadbolt are as follows:

1. Cut the door using the template as a guide (Figure 13.38).
2. Place the wave spring inside the cylinder guard.
3. Insert the cylinder through the cylinder guard with the wave spring *under* the head of the cylinder.
4. Thread the cylinder into the lock body securely. The keyway must be horizontal and on the side of the cylinder closest to the bolt. If the keyway does not come to the horizontal, loosen the cylinder as necessary (always less than one turn).
5. Remove the face plate and install the four set screws at the top and bottom of the lock.
6. Secure the cylinder with the set screws provided.
7. If you have to disassemble the lock, insert a wrench into the hole and turn counterclockwise. This will open the lock.
8. With the lock near or full open, manipulate the lock so the pins slip out of one escutcheon.
9. Reverse this procedure to reassemble. *Note:* The $\frac{3}{16}$-inch roll pin must fit into the hole in the opposite escutcheon for alignment.
10. With the lock assembled and the cover in place, slide the unit into the door cutout (Figure 13.39). Position it so that the longest escutcheon is flush with the high edge of the door bevel.
11. Tighten the lock to the door by inserting an Allen wrench in the appropriate holes and turning clockwise. This brings the escutcheons of the lock together so they clamp the door.
12. Turn the mounting screws finger tight; then give them approximately $\frac{1}{2}$ turn more (Figure 13.40).
13. Install the faceplace with the screws provided (Figure 13.41).

The Ultra 700 lock can be adapted for a single cylinder on the exterior side and a thumbturn on the interior side of the door. For the thumbturn to act as an indicator, the long portion (the actual turning mechanism) must point in the direction of the bolt (extended or retracted position) and the cam must be properly positioned on the thumbturn before the entire assembly is installed in the lock.

Check the cam position. If necessary, remove the drive pin and cam, rotate one half turn (180°), and replace the pin. Figure 13.42 shows the thumbturn as viewed from the interior door side. When installing the lock, be sure the cam is inserted between the mechanism arms; then, tighten the cylinder set screws on the top and bottom of the lock.

The instructions are simple and easy to follow from a locksmith's viewpoint, the time is well spent in terms of customer relations and business income. In addition, if you are replacing the interior cylinder with the thumbturn, you might be able to keep

Figure 13.38 Cut door with template.

Figure 13.39 Slide lock into cutout.

Figure 13.40 Tightening mounting screws.

Figure 13.41 Installing face plate.

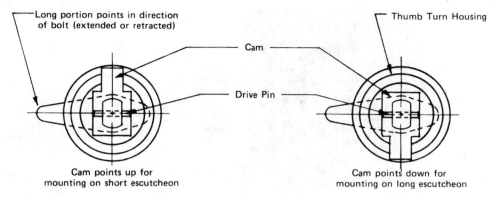

Figure 13.42 Cam direction position when mounting is dependent on whether a short or long escutcheon is used. (*M.A.G.*)

the cylinder that was removed, thus adding various parts to your supply stock. (*Note:* Never resell a used lock; the various individual parts of the lock can be used for repair parts, for study, and for various practical training exercise for new employees.)

M.A.G. Engineering and Mfg., Inc.

The MR series, new from M.A.G., is an adaptation of the original Ultra 700 deadbolt, and one well worth investing in. The adaptation is for use with the Corbin Master Ring cylinder and is now available from locksmithing distributors nationwide.

The adaptation (Figure 13.43) enables locksmiths to use the same masterkeying system but still provides the high-security lock compatible with the present system.

Like the regular Ultra 700, this new series includes no attaching screws on either side of the lock, and the key cannot be withdrawn from the lock unless the locking bolt is either fully extended or fully retracted. Also, the cylinder guard is free-turning and recessed into the escutcheon casting for maximum protection.

A new Series A deadbolt Ultra 700 is also now available for the locksmith to add to potential sales. The Series A is designed for use where one side entrance only is desired (Figure 13.44). The blank side of the lock is a soldering casting. The key side of the lock is available in stainless steel and uses a 1⅛ inch mortise cylinder.

The Ultra 800, another M.A.G. lock, is also a deadbolt, The lock has a $^{15}\!/_{32}$-inch diameter × 1⅛-inch-long mortise cylinder with a standard Yale cam. Figure 13.45 illustrates the lock in an exploded view. Notice the bolt; it is a cylindrical deadbolt instead of the standard rectangular bolt mostly associated with deadbolt locks. Take note of the strike; while it looks like the average mill-run strike, M.A.G. has gone further and ensured that it is ably secured to the frame and supporting stud. In the illustration you can readily see that the interior side of the door will take either a mortise cylinder or thumbturn, providing your customer with an option feature. The lock is set by the factory for 1¾-inch doors, but instructions are included for other door thicknesses, and are also below.

Figure 13.43 The Ultra 700 deadbolt adapted for use with the Corbin Master Ring Cylinder. (*M.A.G. Engineering and Mfg., Inc.*)

Figure 13.44 Ultra 700 deadbolt Series "A" used when one side entrance only is desired. (*M.A.G. Engineering and Mfg., Inc.*)

First, you must determine the proper setting for the lock if the door is not 1¾ inches thick. Figure 13.46 details adjusting the exterior lock housing piece for use on doors of varying thicknesses. The ring referred to is the circular ring on the reverse side (inside portion) of the exterior lock housing. Look at the reverse side, at the 8 o'clock position, to locate the pin hole; insert the pin. With this done, proceed to the door preparation using the template furnished with the lock unit.

Wood doors

Door preparation. Follow the dimensions outlined on the template. Drill ⅛-inch pilot hole through the door face, then drill a 2⅛-inch hole exactly 2¾ inches from the high edge of the door bevel, referring to the template as you do so.

Drill a 1-inch hole through the edge of the door, breaking into the 2⅛-inch hole. Cut out the wood for the bolt housing and face plate 1⅛ inches wide × 2¼ inches high × ³⁄₃₂ inch deep. This face plate hole must be centered on the 1-inch hole.

Drill the four ⅛-inch pilot holes for the four corner mounting bolts. With the pilot holes drilled, replace the ⅛-inch drill bit with a ⁹⁄₁₆-inch bit and put the four ⁹⁄₁₆-inch holes through the door. *Note:* M.A.G. has a drill fixture kit available for fast and ac-

Figure 13.45 Exploded view of the Ultra 800 deadbolt. (*M.A.G. Engineering and Mfg., Inc.*)

curate installation. If you expect to perform numerous installations, then the drill fixture kit is a must-have item for rapid installation.

Prior to mounting the lock, you should assemble the mortise cylinder to the exterior escutcheon of the lock, as follows. (The numbers refer to the various lock part components in Figure 13.45.)

Figure 13.46 Pin adjustment settings for doors with thickness other than 1¾-inch. (*M.A.G. Engineering and Mfg., Inc.*)

Put the cylinder spring (2) over the cylinder (1) then insert the cylinder into the cylinder guard cover (3) and install the escutcheon cover (12).

Thread the cylinder in until the cylinder is tight (the keyway should be toward the rear of the lock and horizontal). Tighten the set screw. The cylinder guard cover should be tight against the escutcheon cover. The ring should rotate freely one-half turn back and forth. If the cam on the cylinder touches inside the mechanism cover, back off the cylinder one turn.

Put the key in the cylinder and rotate one complete turn, first in one direction and then the opposite direction. This is to check the function of this assembly.

If you haven't adjusted the lock for other than a 1¾-99-inch door, remove the mechanism cover by first removing the spring assembly. Pry up the cover and remove the ring; then, adjust the pin for the appropriate door thickness, referring to Figure 13.46 to ensure the details are correct.

Mounting the lock.   Insert the bolt assembly into the 1-inch hole.

Insert the exterior escutcheon (with the mechanism ring in the retracted position) into the ⁹⁄₁₆-inch and 2⅛-inch holes. When inserting, make sure the bolt link fits over the mechanism pin and that the *large tab* of the bolt link is toward the hole in the mechanism cover.

Place the interior escutcheon on the door and secure it to the exterior escutcheon with the four ¼-inch diameter bolts furnished with the lock unit.

Put the cover on the interior escutcheon.

Single cylinder.   Install the thumbturn with the *bolt retracted*. Place the circular wave spring over the thumbturn housing. Be sure the projection on the thumbturn cam and the projection on the housing are together. Place the thumbturn assembly into the interior escutcheon. Make sure the alignment projections line up with the bolt in the escutcheon hole. The thumbturn must be pointing away from the edge of the door. Push the assembly in carefully, making certain the hole in the thumbturn cam lines up with the mechanism pin. Push it in until the catch enters the slot in the thumbturn housing. Check for proper function by rotating the thumbturn and pulling to be sure the catch has engaged the housing. The long side of the thumbturn should point toward the edge of the door when the bolt is extended and away from the door edge when retracted.

Double cylinder.   Assemble the mortise cylinder, cylinder spring, and cylinder guard cover to the interior cylinder housing the same as the exterior escutcheon. The keyway must be horizontal and opposite the catch slot. Thread the interior ring onto the cylinder until it stops against the cylinder housing, then back off one to one and one-half turns prior to assembly into the lock.

Install the interior cylinder housing assembly with the bolt retracted. Insert the assembly into the escutcheon hole, lining up the boss with the slot. Also, line up the ring boss with the escutcheon slot. Press it in fully with the key partially withdrawn, then rotate counterclockwise to stop. Push in on the cylinder until the catch snaps into the slot. Try to turn the cylinder housing clockwise to be sure the catch is engaged to the lock cylinder housing in place. With the key, check for the proper function of the lock from both sides.

Insert the bolt housing assembly into the 1-inch bolt hole. *Caution:* Make certain the opening side of the bolt housing is *up* on left-hand doors and *down* on right-hand doors.

Install the face plate with the screws provided. Make sure the hole in the face plate fits over the projection of the bolt housing assembly to center it properly.

Metal doors. Door preparation and procedures for mounting the lock to hollow metal doors—for doors that have no seam or a hole in the center of the door edge—are as follows.

1. Follow the dimensions outlined on the template. Drill a ⅛-inch pilot hole through the face of the door located 2¾-inches away from the high edge of the door bevel. Be sure the pilot hole is the same distance from the high edge of the door on each side. Bore a 2⅛-inch hole through.

2. Drill four ⁹⁄₁₆-inch holes the same as for wood doors.

3. Using the template, make three centerpunch marks on the edge of the door; one in the center, one ¹³⁄₁₆ inch above, and one ¹³⁄₁₆ inch below center. Drill ¼-inch holes.

4. Open the center hole with a ¾-inch hole saw. Remove the burr on the inside edge. Countersink top and bottom holes for 12-24 flathead machine screws.

5. Install clip nuts to the bolt housing with the long side toward the rear of the housing.

6. Insert the bolt housing through the 2⅛-inch hole in the door and place it against the inside front edge of the door, making certain the projection of the bolt housing assembly fits into the ¾-inch hole. Put the opening on the side of the housing *up* on a left-hand door and *down* on right-hand *doors.*

7. Secure the bolt housing to edge of the door with the two 12-24 screws.

8. Insert the bolt assembly into the bolt housing.

9. Follow steps 2 through 5 for wood doors to complete the installation.

For hollow metal doors that have a seam or fold on the edge that would interfere with drilling of holes for mounting the bolt housing, use this procedure:

Cut out the door edge as shown (Figure 13.47) and mount Install-A-Lock. (Depending on whether the door is a 1⅜-inch or 1¾-inch door, you will require a special order; check with your distributor to see what the Install-A-Lock special order number is for this unit.)

Finish the door preparation for hollow metal doors as previously described. Mount the lock per procedures for mounting to wood doors.

Strike installation

Wood jambs

1. With the Ultra 800 properly installed, close the door fully and throw the bolt out (by either the key or thumbturn) so that the point on the end of the bolt will make a mark on the jamb. This will be the center mark for the strike mounting.

Figure 13.47 Door edge cutout required for metal door with center seam on the door edge. (*M.A.G. Engineering and Mfg., Inc.*)

2. Open the door. Place the strike plate against the door, ensuring that it lines up with the bolt mark centered on the strike plate hole (Figure 13.48).

3. Mark the center hole required to be drilled; also mark around the strike plate edge (Figure 13.49).

4. Drill two 1-inch diameter holes 1⅜ inches deep, one ⅝ inch above and one ⅝ inch below the center mark (Figure 13.50).

5. Chisel out the wood in between.

6. Cut out the wood 1¼ inch wide × 4⅞ inches high × ⅛-inch deep for the strike box and strike plate.

7. Be sure to drill a ⅛-inch pilot hole for all four screws to aid installation.

8. Insert the strike box. Place the finished strike plate over the strike plate box and secure it with screws provided (Figure 13.51).

Figure 13.48 Line up the strike plate to the locking bolt. (*M.A.G. Engineering and Mfg., Inc.*)

Figure 13.49 Mark strike plate. (*M.A.G. Engineering and Mfg., Inc.*)

Figure 13.50 Ensure the midpoint is centered in the middle of the strike plate hole. (*M.A.G. Engineering and Mfg., Inc.*)

Figure 13.51 Strike box into the door frame. (*M.A.G. Engineering and Mfg., Inc.*)

9. Within the strike box, drill two ⅛-inch diameter pilot holes on approximately a 25° angle toward the center of the jamb. This will provide for maximum holding power when the two screws are attached to the stud (Figure 13.52). Figure 13.53 is a side view of the strike box positioning.

Metal jambs not prepared for a strike

1. Mark the jamb with the point on the end of the bolt as described for wood jambs.

2. Drill a ⅞-inch hole only. No strike plate is required.

Figure 13.52 Ensure the screws go into the frame stud for maximum holding power. (*M.A.G. Engineering and Mfg., Inc.*)

Figure 13.53 Side view of the strike box. (*M.A.G. Engineering and Mfg., Inc.*)

3. If desired by the customer, an Adjust-A-Strike may be installed with the unit. Several models are available. Figure 13.54 refers to the models 450/451: 1¼-inch × 2¾-inch lip and deadbolt strike jamb preparation. (Figure 13.55 provides positioning of parts and the strike installation completed for all models described here.)

For 1¾-inch door, start cutout ³⁄₁₆ inch away from the door stop; for 1⅜-inch doors, start the cutout on the edge of the door stop.

1. Cut out 1¼ inches × 2¾ inches, ⁵⁄₃₂ inch deep.
2. Cut out 1½ inches, ³⁄₃₂ inch deep.

For 1¾-inch door, ³¹⁄₃₂ inch; for 1⅜-inch door, ²⁵⁄₃₂ inch.
For latches (lip strike), drill ⅝ inch deep; for deadbolts, drill 1 inch deep.
Install the strike and keepers in the center of the cutout. Close the door to check fit. Adjust if necessary.
For the model 452, for a 1¼-inch × 4⅞-inch lip strike, the jamb preparation is as follows:

1. For 1¾-inch door, start the cutout ³⁄₁₆ inch away from the door stop; for 1⅜-inch door, start cutout on the edge of door stop (Figure 13.56).
2. Cut out 1¼-inch × 4⅞-inch, ⅛-inch deep.
3. Cut out 3⅜ inches, ¹⁄₁₆-inch deep.

For 1¾-inch door, ³⁄₃₂ inch. For 1⅜-inch door, ²⁵⁄₃₂ inch:

1. Install the strike and keepers in the center of the cutout. Close the door and check for proper fit, adjusting if necessary.

The Schlage G Series Lockset

The mechanics of this lockset were discussed in an earlier chapter.

Installation

These instructions, developed from material supplied by Schlage, apply for wood doors from 1⅜ inch to 2 inch thick.

1. Mark the height line (the centerline of the deadlatch) on the door (Figure 13.57). *Suggestion:* Mark the line 38 inches from the floor.
2. Mark the centerline of the door thickness on the door edge.
3. Position the template (supplied with the lockset) on the door with the lower deadlatch hole on the height line. Mark center points for one 2⅛-inch hole and two 1½-inch holes through the template (Figure 13.57). Mark the centers for two 1-inch latch holes on the door edge.
4. Bore one 2⅛-inch hole and two 1½-inch holes in the door panel.
5. Bore two 1-inch holes in the edge of the door. The upper hole should be extended ⅜ inch beyond the far side of the middle hole on the door panel (Figure 13.58). Mortise the edge of the door for the latch front.

STRIKE

SCREW

KEEPER

Figure 13.54 Exploded view illustrating parts relationship for the complete installation of the Adjust-A-Strike. (*M.A.G. Engineering and Mfg., Inc.*)

DOOR STOP

1¼" SEE NOTE NO. 2

SEE NOTE NO. 1

¼" PILOT HOLE 2" DEEP (2 PLACES)

SEE NOTE NO. 4

2⅛"

1¼" SEE NOTE NO. 3

2¾ SEE NOTE NO. 2

CENTER OF LATCH

1" HOLE SEE NOTE NO. 5

Figure 13.55 Jamb preparation for the Adjust-A-Strike models 450 and 451. (*M.A.G. Engineering and Mfg., Inc.*)

Figure 13.56 Jamb preparation and completed installation of Model 452 for a 1¾-inch door. (*M.A.G. Engineering and Mfg., Inc.*)

Mark centers for two 1½ in. holes.

Height Line and Center Line of Lower Deadlatch

Mark center for a 2⅛ in. hole.

Mark centers for two 1 in. holes.

Figure 13.57 Marking holes.

Upper Hole

1½ in. Hole

⅜ in.

1 in. Hole

Over Bore

2⅛ in. Hole

Figure 13.58 Boring holes.

6. Mark vertical and height lines on the jamb exactly opposite the center point of the lower latch hole. Mark a second horizontal centerline 2 inches above the height line for the deadbolt hole. Bore a 1-inch diameter hole 1⅛ inch deep and ¼ inch below the height line. Bore a second hole to the same dimensions ¼ inch above the second horizontal line. Clear out the area between the holes for the strike box (Figure 13.59).

7. Install the strike (it can be reversed for either hand). If you have done the work correctly, the strike screws will be on the same vertical centerline as the latch screws.

8. Disassemble the lock. Remove the inside knob and lift off the outside mechanism (Figure 13.60).

9. With the deadbolt thrown and the crank slot in the vertical position, insert the latch unit and secure it with the screws provided (Figure 13.61). The deadlatch can be rotated to match the door hand (Figure 13.62). The beveled edge of the deadlatch should contact the striker as the door swings shut.

10. Install the inside mechanism with the knob button released and the crank bar in the vertical position. Insert the end of the crank bar in the slot on the deadbolt (Figure 13.63). Engage the jaws of the slide with the deadlatch bar, and engage the deadlatch housing with the ears on the inside mechanism. This sounds more confusing than it is; see the insert for Figure 13.63.

11. From the outside of the door, turn the cylinder bar slot with a screwdriver to check the action of the deadbolt and deadlatch. Once you are satisfied that the lock works properly, retract the deadbolt and release the knob button. Turn the cylinder bar slot to the horizontal position (Figure 13.64). The clutch plate should be situated as shown.

12. Rotate the cylinder bar to the horizontal position with the knob spindle as shown (Figure 13.65).

13. Insert the cylinder bar into the bar slot in the top hole. Engage the clutch plate with the outside knob spindle. The spindle must be positioned as shown in Figure 13.65 and the clutch as shown in Figures 13.64 and 13.66. G85PD locksets require that the cylinder driver bar be flush with the end of the upper bushing (Figure 13.67).

14. Install the inside rose with the turn unit in the vertical position and the V-notch on the inside of the turn pointing to the edge of the door (Figure 13.68). Engage the top of the rose with the mounting plate and, depressing the catch, snap the rose into place. Align the lug on the knob with the slot in the spindle and depress the knob catch. Push the knob home.

15. Test the lock.

Figure 13.59 Clear area between holes.

Figure 13.60 Disassemble lock.

Figure 13.61 Insert latch unit.

Figure 13.62 Rotate deadlatch.

Button
Released.

Engage jaws
of slide with
deadlatch bar.

Engage deadlatch
housing with ears on mechanism

Figure 13.63 Install mechanism.

Cylinder bar
slot should
be horizontal
as shown.

Correct
Clutch Plate
Position

Correct Slide and Latch Contact

Figure 13.64 Testing lock.

Bar must be
horizontal.

Correct
Outside
Spindle
Position

Figure 13.65 Rotate cylinder bar.

Figure 13.66 Positioning clutch.

Correct Clutch Plate Position

upper
bushing

cylinder
driver bar

Figure 13.67 Driver bar must be vertical.

Align turn
with vertical
flats on
bushing with
V-notch
pointing to
edge of door.

V-notch on
inside of
turn.

Depress
rose catch.

Figure 13.68 Install inside rose.

Changing the hand

G series locksets are available in right- and left-hand versions. However, the hand can be changed in the field.

Converting from right-hand to left-hand. Figure 13.69 is an assembled view of a right-hand unit. To convert it to left-hand operation, follow these steps:

1. Disengage the retaining rings (Figure 13.70).
2. Lift off the mounting-plate cover (Figure 13.71).

3. Remove the linkage-bar plate, linkage bar, bolt bar, driver bar, knob-driver linkage arm, and lower bushing (Figure 13.72).

4. With a factory-supplied wrench or long-nosed pliers, straighten the tabs (Figure 13.73).

5. Compress the slide with a screwdriver and pull the assembly out (Figure 13.74).

6. Rotate the spindle 180° to relocate the knob catch (Figure 13.75).

7. Rotate the slide assembly 180°. With the slide compressed, insert the lugs into slots on the mounting plate (Figure 13.76).

8. Bend the tabs to secure the mounting plate (Figure 13.77).

9. Install the linkage arm (Figure 13.78).

10. Install the lower bushing—the slot is vertical (Figure 13.79).

11. Install the retaining ring (Figure 13.80).

12. Install the driver bar—the bar is horizontal (Figure 13.81).

13. Install the bolt bar (Figure 13.82).

14. Install the linkage bar (Figure 13.83).

15. Install the knob driver (Figure 13.84).

16. Install the linkage-bar plate (Figure 13.85).

17. With the upper bushing slot in the horizontal position, install the cylinder driver (Figure 13.86).

18. Replace the mounting-plate cover (Figure 13.87).

19. Install the inside retaining rings (Figure 13.88).

20. Turn the slot in the G85 unit to face the same direction as the slide (Figure 13.89).

21. Turn the deadlatch 180° (Figure 13.90).

22. Check your work (Figure 13.91).

Figure 13.69 Right-hand unit.

OUTSIDE VIEW   INSIDE VIEW

Figure 13.70 Changing hand, Step 1.

Figure 13.71 Changing hand, Step 2.

LINKAGE-
BAR PLATE

LINKAGE
BAR

KNOB
DRIVER
LINKAGE
ARM

BOLT BAR

LOWER
BUSHING

DRIVER
BAR

INSIDE VIEW

Figure 13.72 Changing hand, Step 3.

M504-378 WRENCH

Figure 13.73 Changing hand, Step 4.

SLIDE

Figure 13.74 Changing hand, Step 5.

SPINDLE

KNOB CATCH

Figure 13.75 Changing hand, Step 6.

Figure 13.76 Changing hand, Step 7.

Figure 13.77 Changing hand, Step 8.

LINKAGE ARM

INSIDE VIEW

Figure 13.78 Changing hand, Step 9.

LOWER BUSHING (SLOT MUST BE VERTICAL)

INSIDE VIEW

Figure 13.79 Changing hand, Step 10.

RETAINING RING

OUTSIDE VIEW

Figure 13.80 Changing hand, Step 11.

DRIVER
BAR
(MUST BE
HORIZONTAL)

INSIDE VIEW

Figure 13.81 Changing hand, Step 12.

BOLT
BAR

INSIDE VIEW

Figure 13.82 Changing hand, Step 13.

LINKAGE
BAR

INSIDE VIEW

Figure 13.83 Changing hand, Step 14.

KNOB
DRIVER

R   L

INSIDE VIEW

Figure 13.84 Changing hand,
Step 15.

LINKAGE
BAR
PLATE

R   L

INSIDE VIEW

Figure 13.85 Changing hand,
Step 16.

SLOT
HORIZONTAL IN
UPPER BUSHING

CYLINDER DRIVER
(LUG ENGAGES
BETWEEN LINKAGE-
BAR PLATE AND
KNOB DRIVER)

R   L

Figure 13.86 Changing hand,
Step 17.

INSIDE VIEW

MOUNTING –
PLATE COVER

Figure 13.87 Changing hand, Step 18.

RETAINING
RINGS

INSIDE VIEW

Figure 13.88 Changing hand, Step 19.

SLOT IN
G85 FACE
SAME DIRECTION
AS SLIDE

OUTSIDE VIEW

Figure 13.89 Changing hand, Step 20.

Figure 13.90 Changing hand, Step 21.

SLOT IN UPPER BUSHING HORIZONTAL

SLOT IN LOWER BUSHING VERTICAL

DRIVER BAR HORIZONTAL

LINKAGE - BAR PLATE

BOLT BAR

"L" SIDE OF UNIT

DEPRESS CATCH AND INSTALL OUTSIDE PLATE

Figure 13.91 Changing hand, Step 22.

Converting from left- to right-hand. Figure 13.92 is an assembled view of a left-hand unit. To convert it to right-hand operation, follow these steps:

1. Disengage the retaining rings (Figure 13.93).
2. Lift off the mounting-plate cover (Figure 13.94).
3. Remove the linkage-bar plate, linkage bar, bolt bar, driver bar, knob-driver linkage arm, and lower bushing (Figure 13.95).
4. With a factory-supplied wrench or long-nosed pliers, straighten the tabs (Figure 13.96).
5. Compress the slide with a screwdriver and pull the assembly out (Figure 13.97).
6. Rotate the spindle 180° to relocate the knob catch (Figure 13.98).
7. Rotate the slide assembly 180°. With the slide compressed, insert the lugs into slots on the mounting plate (Figure 13.99).
8. Bend the tabs to secure the mounting plate (Figure 13.100).

9. Install the linkage arm. (Figure 13.101).
10. Install the lower bushing (Figure 13.102).
11. Install the retaining ring (Figure 13.103).
12. Install the driver bar (Figure 13.104).
13. Install the bolt bar (Figure 13.105).
14. Install the linkage bar (Figure 13.106).
15. Install the knob driver (Figure 13.107).
16. Install the linkage-bar plate (Figure 13.108).
17. With the upper bushing slot in the horizontal position, install the cylinder driver (Figure 13.109).
18. Replace the mounting-plate cover (Figure 13.110).
19. Install the inside retaining ring (Figure 13.111).
20. Turn the slot in the G85 unit to face the same direction as the slide (Figure 13.112).
21. Turn the deadlatch 180° (Figure 13.113).
22. Check your work (Figure 13.114).

Attaching and removing knobs

Figure 13.115 illustrates two of the most frequently encountered methods of securing doorknobs. The knob may be threaded over the spindle and held by a set screw, or it may be pinned to the spindle by a screw that passes through the knob and spindle. Another approach is to secure the knob with a retaining lug that extends into a hole in the knob shank. The inside knob can be removed at any time by depressing the retainer; the outside knob can be removed only when the lock is open (Figure 13.116).

Figure 13.92 Assembled left-hand unit.

RETAINING RINGS

OUTSIDE VIEW   **INSIDE VIEW**

Figure 13.93 Right-hand conversion, Step 1.

MOUNTING-PLATE COVER

Figure 13.94 Right-hand conversion, Step 2.

KNOB DRIVER

LINKAGE ARM

LOWER BUSHING

LINKAGE — BAR PLATE

LINKAGE BAR

BOLT BAR

DRIVER BAR

R   L

INSIDE VIEW

Figure 13.95 Right-hand conversion, Step 3.

TABS

M504-378 WRENCH

Figure 13.96 Right-hand conversion, Step 4.

SLIDE

R   L

Figure 13.97 Right-hand conversion, Step 5.

SPINDLE

KNOB CATCH

Figure 13.98 Right-hand conversion, Step 6.

Figure 13.99 Right-hand conversion, Step 7.

Figure 13.100 Right-hand conversion, Step 8.

Figure 13.101 Right-hand conversion, Step 9.

INSIDE VIEW

LOWER
BUSHING
(SLOT MUST
BE VERTICAL)

R   L

INSIDE VIEW

Figure 13.102 Right-hand conversion, Step 10.

RETAINING
RING

OUTSIDE VIEW

Figure 13.103 Right-hand conversion, Step 11.

DRIVER
BAR
(MUST BE
HORIZONTAL)

R   L

INSIDE VIEW

Figure 13.104 Right-hand conversion, Step 12.

BOLT
BAR

Figure 13.105 Right-hand conversion, Step 13.

INSIDE VIEW

LINKAGE
BAR

Figure 13.106 Right-hand conversion, Step 14.

INSIDE VIEW

KNOB
DRIVER

Figure 13.107 Right-hand conversion, Step 15.

INSIDE VIEW

LINKAGE-
BAR
PLATE

INSIDE VIEW

Figure 13.108 Right-hand conversion, Step 16.

SLOT IN
UPPER BUSHING
HORIZONTAL

CYLINDER DRIVER
(LUG ENGAGES
BETWEEN LINKAGE-
BAR PLATE AND
KNOB DRIVER)

INSIDE VIEW

Figure 13.109 Right-hand conversion, Step 17.

MOUNTING-
PLATE COVER

Figure 13.110 Right-hand conversion, Step 18.

RETAINING RINGS

INSIDE VIEW

Figure 13.111 Right-hand conversion, Step 19.

SLOT IN G85 TO FACE SAME DIRECTION AS SLIDE

OUTSIDE VIEW

Figure 13.112 Right-hand conversion, Step 20.

Figure 13.113 Right-hand conversion, Step 21.

LINKAGE-
BAR PLATE

SLOT IN UPPER
BUSHING
HORIZONTAL

BOLT BAR

SLOT IN LOWER
BUSHING
VERTICAL

DRIVER BAR
HORIZONTAL

"R" SIDE OF UNIT

DEPRESS CATCH
AND INSTALL
OUTSIDE PLATE

Figure 13.114 Right-hand conversion, Step 22.

A

B

Figure 13.115 Securing door knobs.

Figure 13.116 Removing knob.

Updating a lockset

The following instructions concern the replacement of a worn or outdated lockset with a new one. The replacement is a heavy-duty G series Schlage, one of the most secure entranceway locksets made. The work is not difficult.

1. Remove the old lockset (Figure 13.117).
2. Remove the latch (Figure 13.118).
3. If a jig is available, use it as a guide for cutting the door; if not, use the template packaged with the lockset (Figures 13.119 and 13.120).
4. Mortise the edge of the door to receive the strikeplate (Figure 13.121).
5. Install the double-locking latch and deadbolt (Figure 13.122).
6. Install the inside lockface assembly (Figure 13.123).
7. Install the internal mechanism.
8. Couple the outside lockface assembly with the internal parts (Figure 13.124).
9. Secure the inside lockface assembly with the screws provided (Figure 13.125).
10. Snap the inside cover into place (Figure 13.126). The lockset is now installed (Figure 13.127).

Figure 13.117 Lockset update, Step 1.

Figure 13.118 Lockset update, Step 2.

Figure 13.119 Lockset update, Step 3.

Figure 13.120 Lockset update, Step 3.

Figure 13.121 Lockset update, Step 4.

Figure 13.122 Lockset update, Step 5.

Figure 13.123 Lockset update, Step 6

Figure 13.124 Lockset update, Step 8.

Figure 13.125   Lockset update, Step 9.

Figure 13.126   Lockset update, Step 10.

Figure 13.127 Lockset update, Step 10.

Strikeplates

Quality locksets are equipped with a deadlocking plunger on the latchbolt. Otherwise, the bolt could be retracted with a knife blade or a strip of celluloid. Nevertheless, it is important to leave very little space between the door edge and the strikeplate. A bolt with a ½-inch throw should have no more than ⅛ inch visible between the door and jamb. If necessary, mount the strikeplate over a steel spacer. This moves the strikeplate closer to the bolt.

The length of the lip is also critical, since a short lip will increase wear on the latchbolt and may frustrate the automatic door-close mechanism, leaving the door unlatched. Mortise strikeplates are mounted on the same vertical centerline as the bolt (or should be). Measure the distance from the centerline of the latchbolt to the edge of the jamb and add ⅛ inch for flat strikeplates and ¼ inch for curved types. Figure 13.128 illustrates a collection of Schlage strikeplates.

10-015

½" Rabbeted Box Strike
For "C": 10-009; For "D": 10-015
Size: 2¾" x 1⅛" x ⅛"
Lips: ⅞" through 3"
     (⅛" increments)
Strike shield may be ordered separately.
Specify finish.

10-003

Standard ½" Rabbeted Box Strike

For: "A", "B", "E"
Size: 2¾" x 1⅛" x 3/32"
Lips: ⅞" through 4"
     (⅛" increments)
Strike shield may be ordered separately.
Specify finish.

10-055

Box Strike

For: B400
Size: 2¾" x 1⅛" x ⅛"
For: B100, B200
10-033
Size: 2¾" x 1⅛" x 3/32"

10-016

Deep Box Strike (¾" Throw) for Fire Door Latchbolt
For "C": 10-010, For "D": 10-016
Size: 2¾" x 1⅛" x ⅛"
Lips: ⅞" through 3"
     (⅛" increments)

10-039

Protected Back Box Strike
For: "A", "B", "D", "E"
10-038 — Lip: 1⅝" for 1⅜" doors
10-039 — Lip: 1¼" for 1¾" doors
Provides added latchbolt protection on pairs of doors.

10-040

Cast Open Back Strike

For: "A", "B", "D", "E"
Size: 2¾" x 1⅛" x ⅛"
Lip: 1" for 1¾" doors
Prevents latchbolt damage on pairs of doors.

10-006

Raised Lip Box Strike
SIZE: 2¾" x 1⅛" x 3/32" thick
LIP: 1¼"

10-004

¼" Radius Rounded Corner Box Strike

For: "A", "B", "E"
Sizes 2¾" x 1⅛" x 3/32"
Lips: ⅞" through 4"
     (⅛" increments)

10-027

¼" Radius Rounded Corner Strike

For: "A" & "F"
Size: 2¼" High
Lip: 1¼" Full

Figure 13.128 Schlage strikeplates.

Entrance-door security

Security is a function of the door, cylinder, bolt, and strikeplate.

- There is *no* security with a hollow wooden door.
- Masterkeying should be kept as simple as possible. Each split pin increases the odds in favor of the lock picker.
- Resist demands for extensive cross keying. These demands originate with customers who insist on single-key performance for executive keys.
- Security begins with proper key-control procedures.
- It is wise to supply extra change-key blanks during the installation phase of the system. Without readily available blanks, the convenient thing to do would be to cut duplicate change keys from master key blanks. Security could be compromised since the change key could fit other locks.
- Double-cylinder locks should be used wherever possible.
- Automatic deadlatches should be specified for all locksets without a deadbolt function. Otherwise the bolt could be retracted by *loiding*, that is, by slipping a flat object between the bolt and the strike.
- Use the longest bolts available to frustrate attempts to gain entry by spreading the door frame. A strikeplate cover can also have the same function.
- Shield the bolts with armored inserts.
- Reinforced strikeplate mounts, particularly with metal door frames, add security by increasing the area of contact between the strikeplate and frame.

Sentry Door Lock Guards

Integral to improved lock security for home, apartment, or business doors is the choice of the door strike. All locks come with a strike that is especially configured for the lock, but you or your customer may wish to have a better strike that will provide a greater amount of security. The Sentry Door Lock Guards Company, of Dania, Florida, has developed a line of strike units that will dramatically improve the security posture of individual door units. The designer and manufacturer of these unique lock guards guarantees that they are jimmyproof.

Perhaps you've never heard of this company or its president, Frank Sushan, but many people and companies have. As the inventor of these unique, impressive, and effective Lock Guard units, he has been profiled in several major newspaper articles. Numerous police departments across the nation now demonstrate the Sentry devices as part of local crime prevention lectures and demonstrations. Further, the National Crime Prevention Institute has approved and endorsed the installation of these products.

Figures 13.129 through 13.133 illustrate the various lock guards available.

LG-1 is a maximum security attachment produced for the do-it-yourselfer who may have no metal cutting tools or experience in security applications. You just remove the present striker and replace it with the fitted unit. The lock guard attachment is installed around the latch face plate on the door when such doors open out.

Figure 13.129 The LG-1 and LG-2 units for door units opening outward. (*Sentry Door Lock Guards Co.*)

Cross section of door Assembly

SECTION
LG-3 Above
For use on DOORS
using METAL JAMBS

Figure 13.130 LG-3 unit is a single, interlocking plate for in-swinging doors hung in a metal frame. (*Sentry Door Lock Guards Co.*)

LG-2 is a single plate that is used with an existing strikeplate for doors that open out. LG-2A is identical to the LG-2, with the exception of the closing section under the T. The WDG-1 cross-section provides the extrusion view of the product.

The LG-3 is a single interlocking door plate for use on inswinging doors that are hung on a metal jamb door frame. Simply, a slot must be cut into the metal door stop to receive the tongue of this interlocking door plate. Any door protected by this unit does not require locking with a key from the outside after the door is closed. The lock cannot be loided, jimmied, or jacked apart. Figure 13.133 shows a cross-section of the installation.

The LG-4 is used similar to the LG-3; when used alone, the difference is that a slot must be made in a wood door jamb to receive the tongue of the unit.

The LG-5 is a two-piece unit for installation on wood doors that open inward only. As indicated for the LG-3 and LG-4, such protected locks cannot be jimmied or jacked open. The doors and jambs will remain together and will not separate.

Figure 13.131 The LG-4 and LG-5 door guard units. (*Sentry Door Lock Guards Co.*)

Figure 13.132 LG-5A and LG-6 lock guards. (*Sentry Door Lock Guards Co.*)

The LG-5A is a replacement safety strikeplate for use on a wood jamb door frame in which the strikeplate screws will not hold the plates onto the jambs, making the purchase of another new lock unit unnecessary. The LG-5B is a replacement safety strikeplate used for the same reason as the LG-5A. This unit provides additional protection where entrance is gained by driving a tool through the molding behind the latchbolt, to force the bolt back to gain unauthorized entrance. The right angle tongue of the strikeplate is extended under the mold for added protection.

LG-6 is an enlarged model of the LG-2 unit, for use on any old or new mortise lock application.

The WDG-1 and WDG-2 are seven foot blank extrusions that are weather strips as well as door and lock guards. They require that all holes are drilled when the unit is installed. The length ensures that it will be long enough to fit any standard door frame, requiring only slight cutting down for perfect fits. Cross-sections of these units are provided in Figure 13.133.

As a point of customer reference, these lock guard units are much stronger and larger than those being used in other applications today. No stronger products are commercially available anywhere.

M.A.G.'s Install-A-Lock

Lock changing and door reinforcement kits provide an extra measure of security for door locks. For old and new doors alike, these units provide the ultimate security protection for the lock and the area immediately around the locking unit.

M.A.G. Engineering and Mfg. Co. has spent many years perfecting these units. Because of the ever-increasing need for upgraded security in today's society, these products are widely used and beneficial to all. Figure 13.134 shows just one type of the Install-A-Lock lock changing kits available.

These kits can be applied to cylindrical locks, deadlocks, deadbolt and key-in-knob locksets, and mortise locksets (Figures 13.135 and 13.136). The lock changing kit enables you to install a lock on any door within minutes. They are made of heavy gauge stainless steel or brass, and the assembled unit brings the ultimate protection to new or old doors; it also covers damaged portions on burglarized doors. Even worn, old, mistreated doors can take on a new glow when graced with an Install-A-Lock kit. Designed to cover jimmy marks, splits and other defacements around old locks, it also provides new strength in the locking area. With the latch plate secured metal-to-metal—*not* wood—the completed installation is extremely solid and secure.

The one-piece construction provides a great savings in labor time. The doors are then protected, after only a few minutes' installation time, right up to the leading edge on both sides.

The security cost is economical and comes in a variety of sizes to meet varying requirements. The cylindrical lock kit will accept locksets with a 2-inch to 2⅛-inch bore and the deadbolt and key-in-knob allow for two backsets of 2⅜ inches or 2⅝ inches. Simply put, these are the finest remodeling kits available.

Figure 13.133 These blank extrusions (top view) are weatherstrips as well as door and lock guards. (*Sentry Door Lock Guards Co.*)

Figure 13.134 Install-A-Lock lock changing kit component parts. (*M.A.G. Engineering and Mfg., Inc.*)

Figure 13.135 Install-A-Lock with a key-in-knob unit, plus a deadbolt locking unit added for extra security. (*M.A.G. Engineering and Mfg., Inc.*)

Figure 13.136 Install-A-Lock for a
mortise lockset. (*M.A.G.
Engineering and Mfg., Inc.*)

Installation instructions

Deadlock or night latch

1. Remove the old lock and all screws from the door.

2. Bore holes for the new lock, if necessary, using the template furnished with the lockset.

3. Chisel out the front edge of the door 1¼ inch wide × 2⅝ inches high × ½ inch deep in the center of the latch hole.

4. Slide the Install-A-Lock unit and position the large hole in the unit to line up with the hole in the door for the lockset.

5. After the unit is in position, install the lock in the door using the 8-32 machine screws that come with the kit. These secure the latch to the unit.

6. Install the four screws and washers to secure the unit to the door. Be sure that the face of the unit is flush with the door face.

7. Install the striker in the door jamb.

For new installations of mortise locks, follow the manufacturer's installation instructions, then follow the instructions below for the Install-A-Lock for mortise locks. For preparing a door in which a mortise lock is currently mounted, follow the instructions below only.

1. Remove the lock from the door.
2. Cut the door edge ⅝ inch deep and 8½ inches high centered on the lock.
3. Install two screw fasteners into the recessed area of the Install-A-Lock with the lip facing down.
4. To determine which side is long, place the unit on a flat surface. The unit will tilt slightly in the direction of the short side. Place the long side on the high edge of the door bevel.
5. Install the lock in the unit with the cylinder at the top.

Sectional trim

1. Select the proper Install-A-Lock insert. Place it between the door and the under side of the Install-A-Lock with the raised area out, fitting it on the large rectangular hole.
2. Install the cylinder and knobs to hold the Install-A-Lock unit in place.
3. Check that the lock matches the strike correctly.
4. Ensure that the Install-A-Lock unit is tight against the door.

Escutcheon trim

1. Place the unit on the door.
2. Install the cylinder and knob assembly to hold the Install-A-Lock unit in place. Do not install the escutcheon retaining screws as yet.
3. Check that the lock matches the strike correctly.
4. Ensure that the unit is tight against the edge of the door.
5. Mark and drill the top mounting hole closest to the edge of the door, using a $^{13}/_{32}$-inch drill, and install through bolt. Repeat the same procedures on the bottom outside hole closest to the door edge; tighten the bolt securely.
6. Drill and install the remaining three through bolts, installing the center bolt first.
7. For escutcheon, drill and install the escutcheon screws as supplied by the lockset manufacturer.

Double-lock units

1. Remove the lock, latch, and all screws (Figure 13.137).
2. See Figure 13.138. Mortise out the front edge of the door (1) 2⅝ inches high × ½-inch deep in the center of the lock hole (2).

3. Using the template furnished with the deadbolt lockset, mark and drill holes (3) and (4) 3⅝ inches from the center of the bottom hole.

4. Slide the unit (5) over the door and position both holes to line up with the holes in the door (Figure 13.139).

5. Install both locksets (6) (before installing the mounting screws). Install both striker plates (7) 3⅝ inches between centers. The strike plate installation should be per the specific instructions for the installed lockset (Figure 13.140).

6. Close the door and check that both the latch and deadbolt (8) enter the strikes (7). If they do not enter, move the Install-A-Lock assembly slightly up or down until the latches enter (Figure 13.141).

7. Install the mounting screws (9) with finishing washers, being sure that the unit is flush with the door.

Figure 13.137 Door section with the old lock and latch removed. (*M.A.G. Engineering and Mfg., Inc.*)

Figure 13.138 Two mortise cuts are required in the door edge in addition to the required holes. (*M.A.G. Engineering and Mfg., Inc.*)

Figure 13.139 Line up the Adjust-A-Lock unit when fitting into the door. (*M.A.G. Engineering and Mfg., Inc.*)

Figure 13.140 Install the locksets and strikeplates. (*M.A.G. Engineering and Mfg., Inc.*)

Figure 13.141 Install the strikes and mounting screws to complete the installation. (*M.A.G. Engineering and Mfg., Inc.*)

Exit Alarm Locks and Panic Bar Deadlocks

Alarmed and deadbolted exit door locks are great concerns of any locksmith's business customers. These units are used to control, monitor, or warn of access (either inward or outward) through various interior and exterior doors. One-way (emergency) doors use these locks. For controlled access areas, such as stockrooms and back rooms, these lock alarm units are a great convenience and additional security.

The Alarm Lock Corporation, which specializes in these devices, has a long list of satisfied customers including American industry, the U.S. government, major chain and department stores, large food markets, factories, hospitals, post offices, schools, and small and medium sized businesses in your own city.

The deadbolt exit alarm lock and the double deadlock panic bar lock evolved to meet a major problem found at government and private-industry buildings. It was (and still is) illegal to lock or bolt an exit or emergency door because it would endanger life. Also, as long as the doors were unlocked, they became an invitation to unauthorized intrusion from the outside. These intrusions led to the theft every year of many millions of dollars from the buildings that were illegally entered.

The solution to the problem is a locking device that allows for emergency exit, sounds an alarm when the door is opened from either the inside or out, and ensures that nobody can enter the area unchallenged. The Safetygard 11 deadbolt alarmed security lock (Figure 13.142) and the Safegard 70 panic bar alarm deadlock (Figure

13.143) are only two of the products that were developed to meet the pressing needs of government and private business.

Safetygard 11 Deadbolt Alarmed Security Lock

The safety alarm lock provides secure, full-time locking of emergency exit doors while complying with standard safety, fire, and local building codes. It provides maximum day and night security and a powerful deterrent against pilferage—from within or without—and always remains panic-proof.

The safety alarm lock provides instant automatic exit in case of an emergency. It combines the best features of panic devices, plus a deadbolt and a warning system. The first person reaching the door applies a slight pressure to the emergency clapper plate, which automatically releases the lock, opens the door, and sounds the alarm that the door is open.

Alarm lock operation. The complete lock mechanism retracts from the edge of the door when the clapper arm depresses, which withdraws the deadbolt and sounds the

Figure 13.142 Safetygard 11 deadbolt alarm security lock. (*Alarm Lock Corporation*)

Figure 13.143 Safegard 70 panic bar alarm deadlock. (*Alarm Lock Corporation*)

powerful twin horn alarm. The alarm continues until the lock is reset to the original position with the appropriate key. The alarm can only be bypassed by an individual authorized to carry and use the bypass key. The mounting plate is already predrilled when the unit arrives from the factory, so it is easily mounted on all styles of doors. The lock is furnace mounted, and it requires no mortising.

Keying. The safety alarm lock unit arrives without the cylinders. (You can sell the cylinder along with the unit, possibly keying the cylinder to meet a certain level of master key operation in medium and large size business and manufacturing complexes.) The safety alarm lock will accommodate all standard rim cylinders. This permits lock integration into an existing or proposed master key system. If special keying or rekeying is not required, the unit should be factory ordered with Alarm Lock's own cylinders; the increased cost is slight, but well worth it.

The lock may be key operated from the outside of the door by mounting an extra rim cylinder through the door. An outside key control cam and cover tamper alarm switch are built into every lock. The outside cylinder, though, must have a flat horizontal tailpiece. The overall length of the cylinder and tailpiece must be a minimum of 3½ inches in length to suit a 1¾-inch door.

Safegard 70 Panic Alarm Deadlock

The Safegard 70 is the world's only available double deadlock panic lock. It provides for anti-pilferage alarm security from within and deadlock security against intrusion from the outside; also, it meets all requirements as an emergency exit device.

With deadbolt locking, the Safegard 70 is designed to deter the unauthorized use of the emergency and fire exit doors. Instant emergency panic exiting is provided, though, as authorized by law. An added advantage is that a remote electrical control for this unit is available.

The unit can be operated in four different modes:

Maximum security locking. Once the door is closed, the key projects the bolt, firmly securing the door by deadbolting it. The crossbar must be depressed to release the deadbolt and deadlatch and open the door. When the crossbar depresses, the alarm activates and remains so until the door is closed and relocked by the authorized relocking key.

Outside key and pull access. By adding an outside cylinder and door pull, you have another mode of access. The key retracts the deadbolt and the pull releases the deadlatch to open the door. Entry remains unrestricted from both sides until the deadbolt is relocked by the key from either the inside or the outside.

Outside key only access. The key must retract the deadbolt and release the deadlatch momentarily to open the door. The door relatches with each closing; the key is required for each re-entry. To relock the door and rearm the alarm, the deadbolt must be projected by use of the key.

Bar dogging function. This provides for free passage from both sides of the door.

The Safegard 70 dimensions are shown in Figure 13.144. Modes of operation for single doors, pairs of doors, and pairs of doors with center mullions are shown in Figure 13.145.

Figure 13.144 Safeguard 70 specifications. (*Alarm Lock Corporation*)

Figure 13.145 Modes of door operation with the Safegard 70 installed. (*Alarm Lock Corporation*)

Narrow stile glass doors

Narrow stile glass doors have evolved the need for a more secure type of locking device than is used in other applications. The architectural trend for such doors will continue for a number of years to come. As an individual concerned with security, you must be extremely aware of what is available to meet a variety of locksmithing situations, both in the home and the general business environment.

Initially, realize that because of this architectural trend, the aluminum used in narrow stile glass doors has shrunk in width and material distance from that of a wooden door with a large piece of glass mounted in it. Because the aluminum is narrower and less thick, it naturally becomes slightly more flexible. At the same time, of course, the space to house the lock also becomes smaller. It is impossible to retract a long bolt in the conventional matter because there is no place for it to go. The conventional lockbolt in the aluminum frame of a glass door has a ½-inch to ⅝-inch throw with an effective penetration of the strike, in some instances, of as little as ¼ inch. This door can be opened by just about anyone with a couple of screwdrivers or, if the frame is weak or worn, with a well-directed frontal attack by a moving body (the intruder). This highly attractive situation means that the lock installed must be able to stop both the hammer and bar and general forced entry.

To an extent, maximum security considerations for sliding glass doors must also be considered. These doors have spawned misconceptions, some fostered by locksmiths who really don't know what they are talking about because they haven't studied the doors, their locks, and the other applications.

This confusion has resulted in some of these doors being installed still lacking the most basic elements of protection that any door should have. Chief among these is a *real* lock—a secure lock that can be operated by a key if desired and one whose key is compatible with the system of other locks in the same building.

Convenience and security are both necessities to the owner/user of the lock, and this is where we now consider the specialized locks that offer the maximum security and protection available for these types of doors. Only one company specializes in locks and locking devices for narrow stile doors; that is the Adams Rite Manufacturing Company of California. The next several pages cover their products.

Deadlocks. The MS1850A deadlock (Figure 13.146) is the first in a series of specialized locks developed especially with the narrow stile door in mind. This lock uses a high bolt of laminated steel nearly 3 inches long and actuated by an uncomplicated pivot mechanism. This lock has made the Maximum Security Deadlock the standard for the entire narrow stile door industry. The length of the bolt provides the maximum security for any single leaf door, even a tall and flexible one. It is also excellent for installation applications where the gap between the door and the jamb is greater than it should be.

The lock case measures 1 inch × 5⁵⁄₁₆ inches × the desired depth. The case depth varies with the backset (Figure 13.147). The cylinder backset can be ⅞ inch, ³¹⁄₃₂ inch, 1⅛ inch, or 1½ inch. Remember that in measuring the backset, it is from the centerline of the face plate to the centerline of the cylinder.

The bolt is ⅝ inch × 1⅜ inch × 2⅞ inches with a 1⅜-inch throw. It is constructed of a tri-ply laminated steel, the center ply having an Alumina-Ceramic core to defeat any hacksaw attack, including rod-type "super" hacksaws.

The face plate for the MS1850A comes in five variations, as illustrated in Figure 13.148. Figure 13.149 shows the dimensions (in inches and millimeters), illustrates the stile preparation for the lock, lock and cylinder installation, and information about the cylinder cam.

Figure 13.146 The MS1850A deadlock for narrow stile doors. (*Adams Rite Mfg. Co.*)

Figure 13.147 Critical lock dimensions for the SM1850A deadlock. (*Adams Rite Mfg. Co.*)

BACKSET	"A"
7/8" 22,2	1.57 39,9
31/32" 24,6	1.63 41,4
1-1/8" 28,6	1.78 45,2
1-1/2" 38,1	2.25 57,1

Figure 13.148 MS1850A face plate variations. (*Adams Rite Mfg. Co.*)

Latch locks. The MS+ 1890 latch/lock (Figure 13.150) is a deadlocked Maximum Security unit offering the highest type of protection after-hours for the businessman or homeowner, in addition to traffic control convenience for business management during the business day.

A typical installation in a bank or a store could require any of three modes of door control

- Both the lock and latchbolts retracted for unrestricted entry and exit during business hours

- Handle-operated latch for exit-only traffic
- Maximum security hookbolt for overnight lockup

This has an operational advantage not necessarily provided by other locks that may seem similar. It requires a 360° turn of the key to throw or retract the hook-shaped deadbolt. When unlocked, turning the key 120° farther will also retract the latchbolt. When using the door handle that can also be obtained for the lock, the handle retracts the spring-loaded latchbolt only.

Figure 13.149 Lock and cylinder preparation, stile preparation, and critical cam dimensions for the MS1850A. (*Adams Rite Mfg. Co.*)

Figure 13.150 MS1890 Maximum Security Latch/Lock unit, with an armored strike and a lever handle. (*Adams Rite Mfg. Co.*)

BACKSET	"A"
31/32"	**1.63** 40,6
1-1/8"	**1.78** 44,4
1-1/2"	**2.25** 57,1

HOW BACKSET IS MEASURED

Figure 13.151 Latch/lock specifications. (*Adams Rite Mfg. Co.*)

Figure 13.151 shows the overall dimensions of the lockset. Figure 13.152 shows the stile preparation, strike location and dimensions necessary, and the latch/lock, cylinders and paddle installation. Latchbolt holdback information is in Figure 13.153.

Figure 13.152 Metal stile jamb and paddle/cylinder installation procedure in an exploded view. (*Adams Rite Mfg. Co.*)

Threshold bolts. Along with a lock installation, you might also be required to install a corresponding threshold bolt mechanism in a narrow stile door (Figure 13.154). The use of the threshold bolt allows maximum security for pairs of doors with the turn of a single operating key. With the simultaneous dropping of a hardened steel hexbolt into the threshold and pivoting of the doors' stile, the two-point lock secures the entire double door entranceway. The threshold bolt is harnessed to the rear of the pivoted bolt. The standard threshold bolt rod is sufficient for a cylinder height up to 53⅞₇ inches maximum. It is fully threaded and can be cut off to allow for low cylinder heights.

A full turn of the key (or a thumbturn if one is installed on the interior side of the door) throws the counterbalanced bolt into the opposite door and the drop bolt into the threshold. The key can only be removed when the bolts are in a positive locked or unlocked position—not halfway. Figure 13.155 shows the dimensions, Figure 13.156 demonstrates the installation of the threshold bolt, and Figure 13.157 illustrates the header bolt installation.

LATCHBOLT HOLDBACK INSTRUCTIONS

THROW
LOCK BOLT

PUSH BACK
LATCH BOLT

3. OPERATE BOLT
HOLDBACK LEVER
WITH KEY, NAIL-
FILE OR SIMILAR
TOOL:

UP TO
HOLD
BACK

DOWN
TO
RELEASE

CYLINDER CAM

CAM
TOLERANCE:
± .005
± ,13

.800
20,32

.120 R
3,05

.182
4,62

MS+ Latch/locks are oper-
able by any standard 1$\frac{5}{32}$"
diameter mortise cylinder
with special M.S.®cam, dimen-
sioned as shown. Cylinders
with M.S.®cams can be readily obtained from most cylinder manufacturers.

Figure 13.153 Latch bolt holdback instruction details. (*Adams Rite Mfg. Co.*)

Figure 13.154 Adams Rite
threshold bolt (left) and header
bolt (right); the unit has the
MS1850A maximum security lock
installed. (*Adams Rite Mfg. Co.*)

Figure 13.155 Threshold bolt and header critical dimensions. (*Adams Rite Mfg. Co.*)

Narrow stile door deadlock strikes. Three basic strikes are available. The majority of Adams Rite deadlocks are installed in metal construction where the strike cutout can simply be a slot. For aesthetic reasons or, in the case of certain locks, for added security, specific strikes should be used.

The *trim strike* (Figure 13.158) is a simple strikeplate that can be surface-mounted on a wood or metal jamb or mortised flush. It uses #10 flathead machine screws and/or two #10 wood screws. It is made of aluminum.

The *box strike* has a dust box added (Figure 13.159). It is customarily used only for wood construction where the dust box prevents chips, sawdust, and other debris from entering the strike.

The *armored strike* (Figure 13.160) has the basic trim plate, but this is backed up by a massive steel doubler designed to prevent the method of forced entry known as *jamb peeling*. It fits within aluminum or other hollow jamb sections; with a trim face flush, the steel is completely hidden.

Figure 13.161 shows mounting methods for narrow stile aluminum doors. The three most prevalent methods of mounting locks and latches within the hollow tube stile of a door are easy.

ANY
BASIC
M.S.
DEAD-
LOCK

INSTALL ADAPTOR ARM IN
LOCK FROM REAR WITH BOLT
RETRACTED AS SHOWN.
DRIVE PIN THROUGH SLOTS
IN LOCK SIDE PLATES PASS-
ING THROUGH HOLE IN THE
ADAPTOR ARM.

ROD/BOLT
LENGTH
MUST BE
APPROXIMATELY
7″ LESS THAN
178   CYLINDER
HEIGHT.

CUT OFF TO
SUIT LOW
CYLINDER
HEIGHTS.

STILE PREPARATION

SEE BASIC M.S.
DEADLOCKS
MS1850A
MS1850 OR
MS1850ANSI

.562
14,3

.69
17,5

.63
16,0

¢ BOLT

.91
23,1

DRILL & C'SK FOR
#10-32 FLAT HEAD
SCREWS ON INTERIOR
FACE OF STILE.

.500
12,7
DIA.
HOLE

THRESHOLD

STRIKE

Figure 13.156 Stile preparation
and installation for the threshold
bolt. (*Adams Rite Mfg. Co.*)

HEADER PREPARATION FOR STRIKE

#6-32 TAP (2) HOLES

.25
6,3

1.00
25,4

1.500
38,1

1.75
44,4

STRIKE

TOP OF DOOR

STILE PREPARATION

.38
9,6

1.500
38,1

STRIKE FOR M.S. PIVOTED BOLT PLUS HOLES FOR 4 SCREWS:

.199 DIA. HOLE
5,05
C'SK 82° x .396
10,06
DIA. (4 PLCS.)

BREAK OFF TAB AFTER INSERTING SCREWS

DRILL & C'SK FOR #10 FLAT HEAD SCREWS (4)

STRIKE CUTOUT FOR M.S. LOCK

.500
12,7

.50
12,7

Figure 13.157 Installation details for the header bolt. (*Adams Rite Mfg. Co.*)

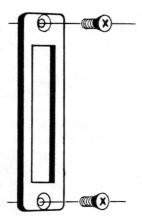

Figure 13.158 Simple trim strike plate for deadlocks in a metal stile door unit. (*Adams Rite Mfg. Co.*)

Figure 13.159 Box strike, customarily for wood frames where the dust box prevents debris from entering the strike. It can also be used on metal door frames. (*Adams Rite Mfg. Co.*)

Figure 13.160 Armored strike unit, essentially used to prevent jamb peeling types of forced entry. (*Adams Rite Mfg. Co.*)

Method A uses the steel bridge spanning the stile. A simple handle, the installation tool, is used to position the bridge accurately in the stile while its two screws are tightened to form a web in the door. Resilient washers allow for minor adjustment to fit the lock flush with the door.

Method B requires two tapped holes in the back web of the stile and tubular spacers of the appropriate lengths. Some door manufacturers use a heavy coil spring in place of the tube to allow for minor adjustments.

Method C is for doors whose stile extrusion shape contains a special inside channel to accept flange clips, on which the lock is mounted. Several door manufacturers have included this channel in their latest designs, so it is something you must be aware of and look for prior to installing the lock. It provides a convenient lock mounting means and adds overall strength to the stile. *Note:* Refer to the Adams Rite Locksmithing Installation Kit, which discusses some of the specific tools necessary for the three methods of mounting locks.

Cylinder pulls.   Cylinder pulls (Figure 13.162) are designed for certain deadlocks and latches. These surface-mounted pulls allow you to key a sliding glass door similarly to other types of doors that have key-in-knob or mortise cylinder locks installed. Dimensions for the cylinder pull are in Figure 13.163 and installation instructions shown in Figure 13.164.

Lock cylinder guards.   The standard mortise cylinder is made of brass and is literally the soft spot in the narrow stile door's security. Using special pliers or other leverage devices, a burglar can tear the cylinder out of the door, leaving an opening through

METHOD "A"

RESILIENT WASHER — 4104 BRIDGE
LOCK MOUNTING PLATE
MOUNTING SCREW

Method "A" spans the stile with a steel bridge at top and bottom of the lock. A simple handle (Adams Rite Installation Tool 4075) is used to position the bridge accurately in the stile while its two screws are tightened to form a "web" in the door. Resilient washers allow for minor adjustment to fit the lock flush in door.

METHOD "B"

TAPPED HOLE
TUBE SPACER
.25
6,35
MOUNTING SCREW
LOCK MOUNTING PLATE

Method "B" requires two tapped holes in the back web of the stile and tubular spacers of appropriate lengths. Some door manufacturers use a heavy coil spring in place of the tube to allow for minor adjustment.

METHOD "C"

FLANGE CLIP 4435
RESILIENT WASHER
LOCK MOUNTING PLATE
MOUNTING SCREW

Method "C" is used with doors whose stile extrusion shape contains a special inside channel to accept flange clips on which the lock is mounted. Several door manufacturers have incorporated this channel in their newest designs. It provides a convenient lock mounting means as well as adding overall strength to the stile.

Figure 13.161   Three methods of mounting a lock in the narrow stile doors using the basic component parts of the Adams Rite installation kit. (*Adams Rite Mfg. Co.*)

Figure 13.162 Cylinder pull set for both sides of a sliding door. (*Adams Rite Mfg. Co.*)

which he can operate the deadlock by hand. The cylinder guard by Adams Rite addresses this problem (Figure 13.165). It offers a three-way defense. First, the outer shield ring offers a poor purchase for either prying or twisting. Second, the ring is of hardened steel so the combination of shape and hardness makes it virtually impossible to grip, even with sharpened tools. Third, in the event that a prying tool, such as a cold chisel, is driven into the stile behind the shield ring, the would-be burglar must pull a heavy steel plate through the round hole in the metal door stile. This degree of leverage is far beyond that available with most hand tools.

The standard package includes the security ring of hardened steel (and free-swiveling when properly installed) and the steel retainer plate to permit the security ring to swivel. The latter is hardened to act as an anti-drill shield. The package also contains spacers for flush fitting in thin stile walls.

Figure 13.166 shows the cylinder guard dimensions. Figure 13.167 provides installation information.

Extracting broken keys. Figure 13.168 illustrates three homemade extractor tools. The first type, made from a length of spring stock, is inserted with the hook down. The round tip of the tool nudges the pins up, and the hook fits into the most forward key cut. The other two are made from *coping* or *fret* saw blades. They are inserted so the teeth snag the edges or top of the key blade.

10-32 TAP (4 HOLES)

SPINDLE

.87
22,1

.43
10,9

2.91
73,9

1.97
50,0

2.55
64,8

.72
18,3

1.25
31,7

.12
3,0

.12
3,0

STANDARD PULLS	
PULL NO.	CYLINDER INCLUDED:
4025	ADAMS RITE
4025-1	NONE (DUMMY)
4025-5	SCHLAGE 22-001
4025-6	WEISER KEYWAY

SPECIAL PULLS	
PULL	FURNISH CYLINDER:
4025-3	LOCKWOOD R-71, H-71, S-71
4025-4	SARGENT SIXLINE
4025-8	CORBIN 471AR
4025-9	RUSSWIN AR848
4025-10	DEXTER 9094
4029	BEST 5A5A2
4025-15	MEDECO 20-800

SPINDLE

TAP 6-32
(2 HOLES)

.975
24,76

1.950
49,53

2.28
57,9

1.22
31,0

.78
19,8

Figure 13.163 Cylinder pull dimensions that can be used with the unit. (*Adams Rite Mfg. Co.*)

1.969
50,01

2.547
64,69

.125
3,17

.375 DIA.
9,52

1848 LOCK

DRILL AND
C'SK FOR
2 #10-32
THRU SCREWS

Figure 13.164 Installation of the cylinder pull; two different pull variations are shown. (*Adams Rite Mfg. Co.*)

Figure 13.165 Hardened steel free-swinging security ring and retainer plate to provide protection against direct attack on the locking unit. (*Adams Rite Mfg. Co.*)

Figure 13.166 Cylinder guard dimensions and positioning when installed. (*Adams Rite Mfg. Co.*)

Figure 13.167 Cylinder guard installation. (*Adams Rite Mfg. Co.*)

Figure 13.168 Three homemade key extractors.

14

Auxiliary Door Locks

The term *auxiliary*, when applied to door locks, has two distinct meanings. It can mean a second lock acting in conjunction with a first and increasing security against intruders. These locks are usually surface-mounted, in which case they are known as *rim locks* or *locking bodies*.

Locks used on interior doors are also known as *auxiliary* locks. Most of these are mortise locks, i.e., they mount inside the door.

Locking Bodies

Locking bodies come in a variety of styles and shapes. A few of the options are shown in Figure 14.1. The bolt may lock automatically when the door is closed. The two locking bodies at the top of Figure 14.1 employ these self-locking latchbolts. Other locking bodies use deadbolts that must be manipulated to lock. Either type may be worked only from the inside or have a keyway on the outside of the door.

Keyless locking bodies with latchbolts have one big disadvantage: they invite lockouts.

Construction

The basic mechanism consists of a case (or cover), a bolt, and a bolt-actuating mechanism. Latchbolts are spring-loaded and are checked by a safety catch. Engaging the safety from the inside puts the bolt and spring in check. The door can be shut without latching.

Locking bodies are adaptable to many different doors. Since they mount on the inside surface of the door, the bevel of the door edge is unimportant. There are times when you will have to choose a particular locking body because of the configuration of the door frame. If at all possible, use the strike intended for this lock because strikes and locks work best as matched teams. If this is not possible, be certain that the strike you use works properly. Figure 14.2 illustrates two of the many strikes available.

Figure 14.1 Assorted rim locks. (*Taylor Lock*)

Figure 14.2 Two strikes by Corbin.

Installation

Every locking body worthy of the name comes with installation instructions, but you will sometimes find yourself working blind. In the absence of manufacturer's instructions, follow this procedure:

1. Position the locking body on the door and scribe its corners and locking body screw holes.
2. Remove the locking body and align the strike on the door frame.
3. Tentatively mark the strike position.
4. Assuming that an outside cylinder will be fitted, drill a hole through the door at the estimated point of contact of the tailpiece (the extension on the back of the cylinder) and the bolt-activating mechanism.
5. Measure the diameter of the cylinder and chuck up the appropriate bit or hole saw. Most cylinders are $1\frac{3}{8}$ inches diameter.
6. Drill the hole from the outside using the first hole as a guide.
7. Break off the splinters and smooth the edges of the hole.
8. Mount the cylinder over the collar and attach it to the retaining plate on the inside of the door.
9. Hold the latch in position and check that the cylinder works properly. You may have to trim the tailpiece.
10. Check the screw holes against the locking body.
11. Start the screw holes with a small nail and drill $\frac{1}{16}$-inch pilot holes $\frac{1}{8}$ inch deep.
12. Mount the locking body and tighten the screws.

It may be necessary to chisel out part of the frame behind the strike. Some are surface-mounted.

The customer may want a set of instructions so he can do the work himself. In this case, the following instructions would be helpful:

1. Drill a cylinder hole in the door $2\frac{3}{8}$ inches from the edge for doors opening inward and $2\frac{1}{2}$ inches for doors opening out.
2. Install the cylinder. Make sure the pins are on top of the keyway.
3. Position the locking body even with the edge of the door and check that the tailpiece operates the latch when the key is used.
4. Attach the locking body.
5. Install the strike.

Inside Door Locksets

The Corbin 7000 series is typical of the better locksets (Figure 14.3). Designed for exterior or interior use, it has these features:

- Case—cast iron and lacquered.
- Deadbolt—forged brass with a ⅝-inch throw. Hardened steel inserts are optional.
- Latchbolt—extruded bass with a ⅝- or ¾-inch throw, depending on application. Antifriction latchbolts are available.
- Latch holdback (safety catch)—optional.
- Hub—cold-forged bronze.
- Strike—wrought brass or bronze with the edge of the lip ⅛ inch from the centerline as standard.

Installation

Installation of Corbin and other locksets was discussed in the previous chapter.

The Taylor 850 (Figure 14.4) is a massive bolt rim deadbolt specifically designed for the extra measure of protection required on doors in a home or business. The bolt measures 1¹¹⁄₁₆ inches high by ⅝ inch thick, which is twice the size for a conventional bolt in a locking unit of this type. In addition to a full 1-inch extension beyond the case, the bolt also possesses two hardened steel inserts that are intended to block sawing the bolt in half.

The strike, like the lock itself, is also massive in design and size and, unlike more conventional strikes, uses five screws set on two face angles to provide an optimum penetration and holding strength once applied.

Figure 14.3 The Corbin 7000 series mortised lockset.

Figure 14.4 A heavy-duty deadlock. (*Taylor Lock Company*)

Figure 14.5 Apartment lock. *(All drawings in this series courtesy Emhart Corporation)*

Figure 14.6 Communicating lock.

The deadlock is operated from the inside by a thumbturn and from the exterior by a key. Normally, the unit is randomly keyed at the factory, but when several are required, they can be keyed alike (KA) or masterkeyed (MK), as circumstances warrant.

Like other units of this general type, lock installation is easy and takes an extremely short time to mount and put into operation.

Variants

Indoor locksets have evolved according to function:

Apartment lock. The latchbolt is released by either knob unless the outside knob is locked by the stop button or deadbolt. The stop button can be released manually at the button or by turning the inside knob. Turning the key retracts the deadbolt; turning the inside knob retracts the deadbolt and the latchbolt (Figure 14.5).

Communicating lock. The latchbolt is released by either knob. A split deadbolt allows operation by either turnpiece (Figure 14.6).

Deadlock. There are three variations to the familiar deadlock: deadbolt operation by key from one side only (Figure 14.7A); deadbolt operation by key from both sides (Figure 14.7B); deadbolt operation by key on one side and turnpiece on the other (Figure 14.7C).

Dormitory lock. The latchbolt is released by either knob, unless the outside knob is locked by the deadbolt. The deadbolt deadlocks the latchbolt. The deadbolt is activated by the inside knob (which also retracts the latchbolt and unlocks the outside knob) by turning the turnpiece or the key (Figure 14.8).

Figure 14.7 Deadlocks. The lock can be controlled by a single key (A), two keys (B), or by a key and a turnpiece (C).

Figure 14.8 Dormitory lock.

Classroom lock. The latchbolt is released by the knob on either side, unless the outside key locks the outside knob. In this mode the inside knob still opens the door (Figure 14.9).

Office lock. There are three variations of the office lock. The first does not have a turnpiece (Figure 14.10A). The latchbolt is released by the key from the outside and either knob, unless the outside knob is locked by the key. Alternately, this same type of lockset may be fitted with a turnpiece on the inside (Figure 14.10B). In this case, the latchbolt is worked by the key on the outside and the knob on either side unless the outside knob is locked by the key. The deadbolt is released by the inside turnpiece and the outside key. The third variation has a key on both sides (Figure 14.10C) and works like the lock just described, except that the key, rather than the inside turnpiece, works the deadbolt.

Storeroom lock. The latchbolt is activated by the knob on either side. The deadbolt operates by the outside key or the inside turnpiece (Figure 14.11).

Figure 14.9 Entrance or classroom lock.

Figure 14.10 Entrance or office locks can have an outside key (A), an outside key and turnpiece (B), or inside and outside keys (C).

Figure 14.11 Entrance or storeroom lock.

Hotel room lock. The deadbolt is activated by the turnpiece on the inside. The latchbolt is retracted by the inside knob and by the guest, master, and grand master keys. The inside knob retracts the deadbolt and latchbolt simultaneously. Only the emergency key can release the deadbolt from the outside (Figure 14.12).

Passage or closet latch. Turning either knob releases the latchbolt (Figure 14.13).

Figure 14.12 Hotel room lock.

Figure 14.13 Passage or closet latch.

Figure 14.14 A privacy lock.

Figure 14.15 Storeroom or closet latch.

Privacy lock. The latchbolt is released by either knob. The deadbolt is activated by the inside turnpiece or by the inside knob (Figure 14.14).

Storeroom lock or closet lock. The latchbolt is released by either knob. The deadbolt is activated by a key on the outside (Figure 14.15).

Hardware

Figure 14.16 illustrates a sample of inside and outside door trim offered by one manufacturer. Figure 14.17 shows five *turnpiece* (*thumb knob* or *turn knob*) options from the hundreds that are available. Many options are purely cosmetic; others are mechanical parts that must be replaced with the exact equivalent. Prime examples are knob and lever spindles (Figure 14.18). You would do well to stock a variety of these parts (Figure 14.19).

Every locksmith, especially a beginner, must have a variety of spindles and knob screws on hand for repair, direct sales, or specific job requirements. With so many different types available, you cannot survive without the variety immediately required. Figure 14.20 shows the technical specifications of the various spindles and knob screws.

These spindles are threaded at 20 threads to the inch. They have drilled and tapped holes at $\frac{9}{24}$ thread and are $\frac{9}{32}$ inch square. In most cases, these spindles can substitute for spindles found in other comparable locks of various brands. There are locks, however, that use spindles with varying threads. Do not attempt to force spindles because this will damage the threads.

The set screws, knob washers, and closet spindles are often used for specific lock installations, and these sizes should be kept on hand, usually in the same storage box with or next to the spindles because they are closely related and used together.

View of inside trim

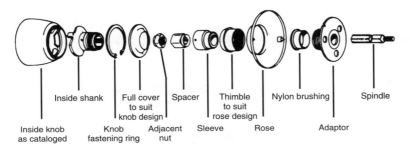

Inside shank   Full cover   Spacer   Thimble   Nylon brushing   Spindle
                to suit   to suit
                knob design   rose design

Inside knob   Knob   Adjacent   Sleeve   Rose   Adaptor
as cataloged   fastening ring   nut

Figure 14.16 Inside and outside trim. (*Emhart Corporation*)

Figure 14.17 Turnpieces by Corbin.

No. 30L

No. 32L

No. 32KL

No. 32LK

No. 130

No. 132

No. 20

No. 30

No. T7055

No. 21

No. 32

No. T7058

No. 22

No. 40 Used with locks operated by knob from one side only

No. 7240

Figure 14.18   Knob and lever spindles by Corbin.

WASHER 128F36-8 FOR 1¾ DOOR

LOCK MORTISE

FOR 1 3/4" DOOR
SPECIFY
    FOR OTHER DOOR THICKNESS

(80034 SERIES)

ADAPTOR PLATE 189F09

VIEW OF OUTSIDE TRIM

PIN 228F54

SPINDLE   NYLON BUSHING   THIMBLE TO SUIT ROSE DESIGN   FILLER TO KNOB   COVER TO SUIT DESIGN   OUTSIDE SHANK   OUTSIDE KNOB AS CATALOGED

ADAPTOR   ROSE AS CATALOGED   SLEEVE SCALP   KNOB FASTENING RING

Figure 14.19   Knob mechanism, exploded view.

STRAIGHT

432....3½"   433... 4"   433A...4½"

434 ...3¼"   435A...4"
435 ...3⅝"   436....4½"

SWIVEL

424....4½"

425....4½" x ⅜" Centers

426....4½"

427....4½" x ⅜" Centers

UNIVERSAL

437....3⅝"   438....4½"

439....4½"

440....4½" x ⅜" Centers

441....2½"

SPLIT

428....2½"

430....2½"

KNOB SCREWS

KSS
Set Screw
¼-28 x 9/32"

SKS
Side Knob
Screw
9-24 x ⅜"

KW
Knob Washer
⅜" I.D.

CLOSET SPINDLES
(Brass Plated)

460....3¼" with rose

461....3¼" with rose

SPECIFICATIONS:

Thread: All threaded spindles, 20 threads to inch. Drilled and tapped holes, 9/24 thread.
Size: All spindles on this page, 9/32" square.
Centers: Center bushings where supplied, ⅜", are removable.
Packing: 1 Dozen per box.
Weight: Average weight, 1 lb. per dozen.

Figure 14.20 Spindle styles and knob screws. Included here are individual unit dimensions allowing you to select a correct spindle to fit a specific lockset. (*Taylor Lock Company*)

Office Locks

Office locks represent an important and lucrative market and one that is easy to overlook. An important part of the security of every business depends on office locks. There is a profusion of manufacturers in this field, making thousands of different locks, each with particular features. Almost all of these locks fall into three groups: rotary cam locks, sliding-bolt locks, and plunger locks.

Rotary Cam Locks

Simplest of office equipment locks, most rotary cam locks employ disc tumblers, although you will occasionally run across one with pin tumblers. The cam is oval and rotates with the cylinder. Those with a 90° throw must be locked for key removal; 180° throw locks can be unkeyed in the locked or unlocked position (Figure 15.1).

The lock body might be threaded (as shown in the illustration) or it might be secured by a retaining clip. The cam might be integral with the lock or secured by a screw and washer. A removable cam is an advantage because it allows the same lock body to be used in a variety of applications. Locks that turn 90° have a single tumbler chamber; locks that turn 180° employ a double chamber and may have a stopping pin, notch, or blocking member at the rear that limits rotation. In some cases, it is possible to change the rotation in the field. The lock may then be used in other applications.

Sliding Bolt Locks

Most sliding bolt locks are disc tumblers, although pin tumblers are not unknown. There is no cam; the bolt is grooved to accept a projection on the back of the plug (Figure 15.2). The projection engages the groove and converts the rotary motion of the plug into reciprocating (to and fro) motion. The bolt is generally heavier and more durable than those of other office locks. This arrangement allows the key to rotate 180°; the key can be withdrawn in either the locked or unlocked position.

Figure 15.1 A Medeco high security cam lock. (*Medeco Security Locks*)

Figure 15.2 A Corbin sliding bolt lock, widely used on cabinets and drawers.

The most serious sliding-bolt lock is known as the *T-bolt* because of its shape (Figure 15.3). The locks almost always employ a disc tumbler cylinder with the bolt-actuating pin cast as part of the plug. The plug can generally be released with a probe wire.

If the key is not available and the bolt is thrown, you have three options. The best is to pick the lock. Once the drawer is open, removing the retainer is no problem. Another approach is to drill through the drawer. Figure 15-4 shows the approximate drilling point. Drill no deeper than ⅞ inch to avoid damage to the lock itself. Finally, you can drill out the plug, destroying the lock in the process.

Plunger Locks

Plunger locks are usually found on sliding cabinet doors. These locks are mounted on the outside door and the bolt, extending from the rear of the plug, engages a hole in the other door (Figure 15.5). The doors are locked against each other so neither can be moved. A spring returns the plug to the open position when the key is turned.

Although disc tumbler types were once popular, pin tumblers are almost universal today. Locksmiths new to the trade might never have seen a disc-plunger lock.

The plug is usually secured by a screw located at the side of the cylinder and set in a notch in the lock (Figure 15.5). The screw serves a double purpose: it determines the throw of the bolt and aligns the plug with the cylinder.

Figure 15.3 A Corbin T-bolt lock, generally stronger than other sliding bolt locks.

Figure 15.4 Occasionally it's necessary to drill a desk lock for access to the retainer clip. A ¹⁄₁₆-inch hole, ⅞ inch deep or less, is adequate.

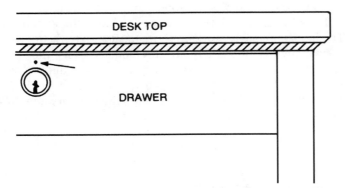

Figure 15.5 A Corbin plunger lock for securing sliding doors.

The bolt is a projection from the rear of the plug unit. Without the key you can pick the lock or force the doors far enough apart to disengage the bolt. The latter method is the faster of the two.

Filing Cabinet Locks

Filing cabinet locks are fairly standardized and use a bolt extending from the top of the lock body (Figure 15.6). The bolt manipulates the control bars that lock each drawer of the cabinet.

The bolt is spring-driven; unless the plug is turned, the bolt extends out of the locking body. In other words, the drawers are automatically locked. The bolt may be round or rectangular in cross-section. The lower face of the bolt—the part buried in the cylinder—is recessed for the spring. The plug cam works in a notch on the side of the bolt.

Open keyways (as opposed to *blind* or *masked*) are generally used. The keyway runs from the face through the plug body. You can open the lock with relative ease. Use the tool shown in Figure 15.7 to work the bolt directly.

During the past decade, new standards have been set for filing cabinet locks; many manufacturers have revised their dies to do away with the open keyway and the breach in security it represents. Today, many filing-cabinet-lock bolts are isolated from the keyway. A few manufacturers use their original dies but add a fifth pin. This pin is longer than the others and has no upper chamber. It cannot be lifted, so it blocks access to the bolt. Another option has been to block the keyway end with a horizontal pin. The jimmy no longer works, and we must turn to the more traditional methods of opening a keyless lock. There are three choices: pick the lock, drill the plug, or force the bolt. Picking is the most desirable of these alternatives, but requires patience and skill. Drilling is fast and means the loss of the lock, but it is preferable to forcing the bolt. If the customer accepts the risk, pry one drawer open far enough to manipulate the bolt with a letter opener.

Figure 15.6 A filing cabinet lock in cross-section.

Figure 15.7 A jimmy for open-keyway locks. The tool is made of spring steel, ⅛ inch to ³⁄₁₆ inch wide, and slips through the keyway for direct manipulation of the bolt.

16

Automobile Locks

While the automobile has been with us since the beginning of the century, the lock was adopted slowly. However, by the late 1920s nearly every auto had an ignition lock, and closed cars had door locks as well. Current models can be secured with half a dozen locks.

Sidebar Locks

While automobiles can be fitted with pin or disc tumbler cylinders, the sidebar lock is most typical and the only one that merits special attention. The other two were discussed earlier.

Figure 16.1 illustrates this lock in simplified form. Notice that the wafers have V-shaped notches in their sides. Inserting the correct keys aligns the notches, and the spring-loaded sidebar moves out of the cylinder and into the plug. Once the sidebar passes the shear line, the plug is free to rotate.

Key fitting

Key fitting for disc and pin tumbler locks was described in earlier chapters. The techniques used for automotive variations of these locks are no different.

Key fitting sidebar locks requires disassembling the lock. Remove the lock from the vehicle and, with a small screwdriver or other sharp tool, release the staking that holds the sidebar and associated spring in place.

Remove the tumblers and their springs. Insert the appropriate blank into the keyway and replace the tumbler at the far end of the plug. Working from the sidebar chamber, scribe a mark on the blank where the blank touches the tumbler. This is the guide mark for the cut. File the mark lightly and reinsert the blank; the file mark should be at or under the tumbler. File slowly until the V-notch on the side of the tumbler appears to align with the sidebar. Hold the sidebar and spring temporarily in

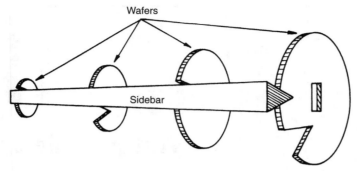

Figure 16.1 How a sidebar fits into tumbler cutouts.

place with your fingers and try the key. If all is right, the sidebar will extend into the notch, freeing the plug. The key should turn.

Repeat this operation for each of the remaining tumblers.

Door Locks

Modern door locks are disassembled from inside the door. Figure 16.2 shows a typical example. Remove the arm rest (7 in the drawing) and inside handles. Most handles are secured by a spring clip that is accessible with the tools shown in Figure 16.3. Note the positions of the handles on their spindles for assembly.

Remove the Phillips screws holding the door trim (6) to the door body (2). Gently pry the trim away from the body and set it aside. The door lock mechanism is visible through access holes in the inner door-body panel.

Key to Fig. 21-2

(1) Door hinge (upper)
(2) Door body
(3) Damper
(4) Door hinge (lower)
(5) Door weatherstrip
(6) Door trim
(7) Arm rest
(8) Door glass
(9) Main roller guide (front)
(10) Channel
(11) Regulator assembly
(12) Regulator handle
(13) Upper stopper (lower)
(14) Sub-roller guide
(15) Upper stopper
(16) Main roller guide (rear)
(17) Door inside handle
(18) Door inside lock
(19) Door outside handle
(20) Cylinder lock
(21) Door lock
(22) Door striker
(23) Shim

Figure 16.2 A typical front door assembly. (*Chrysler Corp.*)

Figure 16.3 These K-D door handle tools are available from auto parts stores. View A illustrates the tool for Chrysler retainers; view B, the tool for Ford and General Motors.

Most problems involve the lock mechanism and not the cylinder. Check for broken springs, missing retainer clips, and rust on working parts. The cylinder (20) is secured by a U-shaped retainer clip. Cylinders can be disassembled, although modern practice is to replace the cylinder as a unit rather than attempt repairs.

Move the striker to align the door with the body panels and compensate for door sag. Loosen the mounting screws a quarter turn or less and, using a soft mallet, tap the striker into position (Figure 16.4). Shims are available from the dealer to move the striker laterally out from the door post.

Figure 16.4 Door strikes are adjustable. (*Chrysler Corp.*)

Figure 16.5 The door adjustment is at the pillar side of the hinges for most vehicles. (*Chrysler Corp.*)

While this is out of the province of locksmithing, as such, it is useful to know that the door can be adjusted at the hinge on all automobiles except the Vega (Figure 16.5). Loosen the pillar-side bolts slightly and manipulate the door.

Ignition Locks

Early production ignition locks are secured to the dash by bezel nuts or retainer clips and feature a *poke hole* on the lock face or in the back of the cylinder that gives access to the retaining ring. In a few cases, screws hold the lock from the underside. Late-model locks (the kind that lock the steering wheel as well as the ignition) are not something that the average locksmith should become involved with. These locks can be serviced but special tools and special knowledge are required. You must dismantle the upper portion of the steering column and, in some cases, drop the steering column from the dash hanger. A mistake in assembly—even using the wrong screw—can defeat the collapsible steering column. If you want to get into this aspect of the trade, spend a few weeks at dealer schools and invest in the proper tools.

Glove Compartment Locks

Modern glove-compartment locks are typically secured by a retainer accessible from the front of the lock. Figure 16.6 shows how this is done. Make a ¼-inch hook in the end of a piece of Bowden (hood-latch) cable. Insert the hook into the keyway and work the retainer free. Most are installed with the open ends toward the passenger door; nudge the retainer toward the center of the car to disengage it.

Extract the cylinder with the key and release the bolt by hand. The cylinder is stamped with a numerical keying code—the same code that is used for the trunk or, in the case of station wagons, the tailgate.

Trunk Locks

Figure 16.7 shows a luggage-compartment latch typical of those found on middle-priced automobiles. More expensive cars use a more complex lock mechanism that may be solenoid-tripped from inside the vehicle. The basic parts and their relationships are the same. The cylinder is secured to the trunk lid by a retaining clip and works the lock assembly through a tailpiece or extension. (A few cars mount the cylinder on the lower body panel and the striker on the lid.) The striker is located on the aft body panel.

Most lock assemblies are adjusted only at the striker plate. The plate is mounted on elongated bolt holes and can be moved up, down, left, and right. Other types have an additional adjustment at the latch. The idea is to position the parts so the lid closes without interference from the striker and tightly enough to make a waterproof seal between its lower edge and the weatherstripping.

Figure 16.6 The glove box cylinder can be removed from the outside with the tool shown. (*Ford Motor Co.*)

Figure 16.7 The trunk lock assembly used on Mustang and Cougar automobiles. (*Ford Motor Co.*)

Decoding, Picking, and Emergency Entry

A valuable locksmithing skill is the ability to cut different key types when only the lock manufacturer and key number are known. The locksmith might never see the lock or key; at best she might have only the lock. Certain information might be required of the customer, depending on city or county ordinances. In most cases, the locksmith will need to find out who owns the lock. Has the owner given authorization to have a key made for the lock? Requests involving high-security locks must be forwarded to the factory together with a written release from the owner. The code varies with each manufacturer, but includes the key blank type, cut spacing, and cut depths.

Reading Codes

Direct-digit and indirect codes are most popular. The *direct-digit code* consists of one or two letters followed by a series of numbers. The letters identify the lock series, and the numbers refer to the depths of the cuts. The series might start at the last tumbler (the one closest to the key tip) or at the first (next to the key shoulder). *Indirect codes* are safer; the numbers must be decoded before the cuts can be identified. Professional locksmiths have the necessary code books.

The smallest number—either 0 or 1—in both kinds of codes represents the shallowest cut; 9, the last of the series, represents the deepest cut. Once you have the number, you can usually determine the type of lock. For example, key code LL76 is limited to desks. This fact can be useful in that it narrows the range of possible blanks.

The code book says that LL76 is a Yale desk lock and that you can use ILCO 01122A, National Y12, or Yale 9278A blanks. The depth and spacing charts show five depths, each 0.020 inch apart, and five spacings. Counting from the tip to the shoulder, the first four cuts are spaced 0.094 inch apart; the last cut is 0.140 inch from the shoulder. Code number LL76 translates as 24253, and it is read from the tip to the shoulder. To cut a key with this code, place a Yale No. 2 depth and spacing key in the vise together with the appropriate blank. Cut the No. 2 bits. Repeat the operation with Nos. 3, 4, and 5 depth and spacing keys.

Depth and spacing keys

Depth and spacing keys simplify the work. These keys are milled to various depths and spacings. Because manufacturers vary these dimensions, you need a complete key set or at least one that covers the major American locks.

Schlage wafer-tumbler keys

Schlage keys are easy to cut by code because the cuts all have the same depths. Three Schlage keys and one depth key are required.

Lock picking

Lock picking is the ultimate test of a locksmith; it tries his knowledge as well as his skill. To many customers, lock picking is the one ability that distinguishes a true locksmith from someone who merely tinkers with locks.

Lock picking tools

Basic lock picks are illustrated in Figure 17.1 and include the following:

- Half-round feeler
- Round feeler
- Rake
- Half-rake
- Diamond
- Double-diamond
- Circular
- Reader
- Extractor
- Mailbox
- Flat lever

You can make your own lock picks from flat, cold-rolled steel 0.020 to 0.025 inch thick. The strip should be about 6 inches long and at least ⅛ inch wide. One end may be fitted with a handle, or both ends may carry working surfaces.

Picks for warded locks come in various designs. A few of them are shown in Figure 17.2. It's usually a simple matter to make one of these picks. Sometimes a standard precut key shaved to pass the keyhole wards is all that is needed.

HALF-ROUND FEELER

ROUND FEELER

RAKE

HALF-RAKE

DIAMOND

DOUBLE-DIAMOND

CIRCULAR

READER

EXTRACTOR

MAILBOX

FLAT LEVER

Figure 17.1 Basic lock pick shapes.

Tension tools for use with various picks are made of hardened spring steel. Normally they are from 3 to 5 inches in length, excluding the parts that actually fit into the keyway opening.

A letter box tension tool is made from a thin strip of spring steel about 4 inches long and has a ⅛-inch to ¼-inch probe at its working end. It is best to have several with different probe lengths to ensure that you always have the correct one on hand.

Picks and tension tools for double-sided disc tumbler locks are unique in their design (Figure 17.3). They should be purchased from your locksmith supply house; it's cheaper to buy them than to make them.

Figure 17.2 Picks for warded padlocks.

Figure 17.3 Picks for double-sided disc tumbler locks.

Lever lock picks come in two varieties—those for true lever locks and those that double as key picks. The true lever lock pick is made from 0.095-inch or thinner spring steel and it has a tension tool that is the same thickness (Figure 17.4). The second variety is cut from spring steel or it is made from standard key blanks.

Note: In many jurisdictions, it is illegal to possess lock picks without a locksmith's license. The penalties are severe, even for first offenders.

Lock picking techniques

The prerequisite for successfully picking a lock is a sound knowledge of how the lock mechanism works. Without such an understanding, picking a lock becomes a matter of sheer luck.

Picking the pin tumbler lock. To pick a pin tumbler lock you must be able to determine when the cylinder pins are at the shear line. Apply just the right amount of pressure on the cylinder core with a tension tool. Figure 17.5 shows a feeler pick raising a pin to the shear line. To raise the pins in this way requires constant practice. Too much or too little pressure on the pick will cause the pin to stick at the wrong place (Figure 17.6).

Factory precision in drilling the cylinder pin holes is never perfect. An enlarged view of the top of a plug would show that the holes are not really in alignment (Figure 17.7). This is helpful in opening the lock. Since the holes are not perfectly aligned, you can hold one pin at the shear line while working on others. Another built-in advantage is the fact that locks always have at least some clearance between moving parts—and a little working space is all a good locksmith needs to pick a lock.

In raising the various pins, it is preferable to start with those of the greatest length. By using this method, you progress from the smallest amount of pick movement up to the greatest. As each pin reaches the shear line, the core moves ever so slightly; you will be able to feel this movement when it is transferred to your tension tool.

Using a feeler pick, move one pin at a time. Since the purpose of the pick is to move the individual pins up *to* the shear line, not *above* it, you must take your time. If you feel that all but one of the pins have reached the shear line and the last one

A

0.095 in.
SPRING
STEEL
(FLATTENED
AT
POINT)

B

Figure 17.4 A tension tool (A) and picks for warded locks (B).

FEELER PICK

Figure 17.5 A feeler pick raising a pin to the shear line. (*Desert Publications*)

Figure 17.6 Too much pressure with a pick will raise a pin above the shear line.

Figure 17.7 Cylinder pin holes are never in perfect alignment.

just won't go, you have probably pushed the last pin above the shear line. Release the tension and start again with slightly less tension than before.

Remember that you must vary the amount of pressure you apply to the pick depending on the amount of resistance you receive from each pin. Also, different locks require different amounts of pressure. What works on one lock will not necessarily work on the next. Even two locks of the same make and style may not require the same amount of pressure.

It is sometimes possible to use picks (especially the rakes) to *bounce* the pins to the shear line. To do this, insert the pick fully then withdraw it quickly while holding light tension on the plug. Simply forcing the pick rapidly in and out of the cylinder won't do the job; it merely bounces the pins above the shear line.

Picking warded locks. Warded locks are very easy to pick. Sometimes you can even use a pair of wires: one for throwing the bolt and another for adjusting the lock mechanism to the proper height for bolt movement. Here, as with any other kind of lock picking, it is a matter of practice on a variety of locks.

Warded padlocks have only one or two obstacles, thus simplifying your choice of pick. Some warded locks (such as the Master) that have keyways with corrugated cross sections can be opened in no time at all by having several precut blanks available. The regular picks are usually just a fraction of an inch too wide to fit into the keyway.

Smaller locks with correspondingly smaller keyways require that you either get a second set of tools and cut them down to size or use several blanks to obtain the proper picks for these locks.

Picking lever tumbler locks. Figure 17.8 shows a lock with the frontplate removed, indicating the position of the tension tool in relation to the bolt. Holding the levers with the tension tool (as shown in the side view) enables you to manipulate the levers with the pick.

In learning to pick the lever lock, it is best to start with a lock with only one lever in place. At first, work on the lock with the face plate removed so you can get an idea of how much pressure to apply to the bolt and how much movement is required of the pick to move the lever into position for the bolt to move through the lever gating.

As you progress to locks with several levers, keep the face plate on. When you encounter a problem, remove the face plate so you can see what you're doing wrong. Always remember to insert the tension tool first. Push it to the lowest point within the keyway so the pick will have maximum working space.

As with the pin tumbler lock, one tumbler tends to take up most of the tension; work on this one first. As you move this tumbler slowly up to the proper position for the bolt to pass through the gate, you will feel a slight slackening in tension from the bolt as it attempts to force its way into the gating. This tension will be transmitted to you through the tension tool. Stop at this point and do the same to the next lever with the greatest amount of tension. Continue until all the levers reach this point. Then, by shifting the tension tool against the bolt, the bolt will pass through the gate and open the lock.

There must be at least some pressure on the tension tool at all times. A lack of pressure will cause any levers in position to drop back into their original locations.

Picking disc tumbler locks. Standard disc tumbler locks (those using a single-bitted key) take the same picks as pin tumbler locks. Also like the pin tumbler lock, a disc

Figure 17.8 A tension tool exerting pressure on lever tumblers.

tumbler lock can be picked by bouncing the tumblers to the shear line. Usually a rake is the best tool for this job, but other picks can be used too.

To develop your proficiency, try opening the disc tumbler lock with the feeler pick, working on each tumbler individually. This helps you learn which picks you are most comfortable with and which ones you prefer for different types of locks.

When working on double-bitted disc tumbler locks, the bounce method is best. Insert the tension tool into the keyway and apply a slight pressure to the core as you pull the pick out of the lock. These locks can also be opened with a standard pick set and tension tools, but it can take from ten minutes to half an hour.

Lock picking considerations

The following are points to remember when picking any lock:

- To give yourself maximum working space, use the narrowest pick available
- Hold the pick as you would a pen or pencil. Don't use wrist action; your fingers work much better in manipulating the pick in the lock.
- Steady your hand with your little finger against the door. When working on a key-in-knob cylinder, steady your hand against the edge of the knob.
- Your pick should enter the keyway above the tension tool without moving any of the tumblers. If it doesn't, the tension tool is either too high in the lock or the keyway grooves are such that the tension tool must go at the very top of the keyway.
- Worn cylinders and loose plugs frequently open easier and quicker with the bounce method—normally within three to five tries.

Emergency Entry Procedures

Sometimes locksmiths gain entry into locked premises by picking open the lock. A good locksmith can also gain entry through special techniques called *Emergency Entry Procedures* (EEP), *special techniques*, *emergency measures*, or *quick entry*. When a lock resists picking, the EEP are used.

Much of the information and illustrations in this chapter come from Desert Publications' book *Lock Out*.

As a professional locksmith, you should have the proper equipment to make emergency entries as quickly and neatly as possible. Some entry tools are manufactured; you should purchase a set of these for your use. Some tools you can fashion yourself. Experience will show you which tools and techniques work best for you.

Using Emergency Entry Procedures can account for a substantial percentage of a locksmith's income; thus, it is wise to be knowledgeable in a variety of entry techniques. It might not seem like good locksmithing practice to force a lock open, but economics, time, and other requirements may make it necessary.

Several preliminaries should be undertaken when preparing to make an entry using EEP. Try to obtain as much information as possible. With this information, you can decide what methods will work. Ask questions about the type of lock, the key numbers, nearby open windows, a possible extra key held by someone else, the condition of the lock, alarms and other devices attached to the door, etc. All of this information will aid your preparations for the entry. Fortunately, most lockouts require only that you pick the lock or make a new key.

Drilling pin tumbler locks

Drill a lock only as a last resort. There are two drilling methods used with pin tumbler locks. The first method is the drilling of the cylinder plug. This has the advantage of saving the cylinder itself. The inner core can be replaced. This method destroys the lower set of pins just below the shear line, allowing the plug to be turned. Follow these procedures to drill the cylinder plug:

1. Drill the plug below the shear line.
2. Insert a key blank and a wire through the drill hole to keep the upper pins above the shear line and the destroyed pins below it.
3. Turn the cylinder core and open the lock.

Once the door is open, dismantle the lock; remove and replace the core and fit new pins to the lock. If you used the plug follower when removing the core, you will only have to fit new lower pins to the lock to match a key.

The second method involves drilling just above the shear line into the upper pins.

1. First, insert a key blank into the lock to force all the pins to the upper parts of their chambers.
2. Drill about ⅛ inch above the shear line or shoulder of the plug and directly above the top of the keyway (Figure 17.9). Use a drilling jig to ensure drilling at the proper point on the lockface. Use a ⅛-inch or ³⁄₃₂-inch drill.
3. Poke a thin wire or needle into the hole and withdraw the key to just inside the keyway. Then, use the tip of the key to turn the core. Removing the key partway allows the bottom pins to drop below the shear line while the wire or needle keeps the upper pins above the shear line.

Once the door is open, dismantle the lock, remove the cylinder, and replace it with a new one.

Figure 17.9 A cylinder removal clamp. (*Desert Publications*)

Cylinder removal

Sometimes the cylinder must be removed to open a lock. This means shearing off the screws that hold the cylinder in place. Several different rim cylinders removal tools are available. The one in Figure 17.10 can be made in your own shop. It's called a *cylinder removal clamp*. To make one, follow these steps:

1. Use a piece of steel tubing about ⅛-inch larger in diameter than the cylinder. The tubing should be no more than 2½ inches long.
2. Cut into the tubing about ¹³⁄₁₆ inch, as shown in Figure 17.10.
3. Cut a center hole, ¼ inch to ⅛ inch in diameter, at the end of both cuts and insert a steel rod about 10 inches long. The rod becomes your handle for turning the cylinder.
4. Drill two holes in the tubing so you can insert a bolt perpendicular to the handle. Ensure that the bolt will be ½ inch in front of the handle. The bolt threads must extend on both sides of the tubing.

To remove the cylinder, follow this procedure:

1. After first removing the cylinder rim collar, set the removal clamp over the edge of the cylinder. Tighten down the tension bolt; this provides the gripping pressure to the cylinder.
2. Twist the cylinder clamp by the handles and force the cylinder to rotate, shearing off the retaining screw ends.
3. Remove the cylinder from the lock. When removed, reach in and open the door by reversing the bolt.

You can also use the standard Stillson wrench to remove the cylinder.

Cylinders can also be pulled out of the lock unit, but this method ruins the cylinder threads, requiring you to replace the entire unit. The tool used is called a *nutcracker*. The sharp pincer points are pushed in behind the front of the cylinder face. Clamp down and pull the cylinder out. Many times an unskilled locksmith can ruin

Figure 17.10 A cylinder removable clamp. (*Desert Publications*)

the entire lock and a portion of the door by not knowing how to use this tool properly.

Some cylinders have to be drilled. Follow this procedure:

1. Drill two ³⁄₁₆-inch diameter holes about ⅞ inch to 1⅛ inches apart on the face of the cylinder.

2. Drive two heavy bolts into these holes so that at least 1½ inches are sticking out.

3. Place a pry bar or heavy screwdriver between the bolts and use it as you would a wrench to force the cylinder to turn, shearing off the long screws.

Window entrances

Window entrances are relatively easy. The old butter knife trick is usually successful. Since most window latches are located between the upper and lower windows, sliding a knife up between the windows allows you to open the latch.

Should the area be too narrow for a knife, shim, or other device, drill a ¹⁄₁₆-inch hole at an angle through the wood molding to the base of the catch. Insert a stiff wire and push back the latch.

Office locks

Most office equipment can be opened using a few basic methods and a handful of tools.

Filing cabinets. Though most filing cabinets have the lock in essentially the same position, the locking bar arrangement for the various drawers will vary. These variations have to be considered in working on the locks.

One method is to work directly on the lock itself. Slide a thin strip of spring steel ⅛-inch thick into the keyway, and pull the bolt downward. This will unlock the lock (Figure 17.11).

If the lock has a piece of metal or pin blocking access to the locking bolt, use a piece of stiff wire with one end turned 90°. Insert the wire between the drawer and the cabinet face and force the bolt down with the wire. This will allow you to open the drawers.

If there is not enough room for you to work with the wire, use a piece of thin steel or a small screwdriver to pry back the drawer from the cabinet face to allow you to see and work with the wire.

You can also open the drawers individually if necessary.

1. Use a thin piece of spring steel or a wedge to spread the drawer slightly away from the edge.

2. As your opening tool, use another strip of steel about 18 inches long, ½ inch to 1 inch wide, and 0.020 inch thick.

3. Insert the opening tool between the drawer catch and the bolt mechanism (Figure 17.12).

4. With a healthy yank, pull the drawer open. The opening tool creates a bridge for the drawer catch to ride on and pass the bolt.

A

B

Figure 17.11 Sometimes a thin strip of spring steel inserted into the keyway of a filing cabinet lock (A) can be used to pull the bolt down (B). (*Desert Publications*)

A

B

C

WEDGE

Figure 17.12 Sometimes the drawer of a filing cabinet (A) can be opened by wedging the drawer away from the cabinet frame (B) and inserting a strip of steel between the drawer catch and the bolt mechanism (C). (*Desert Publications*)

Other filing cabinets can be inverted to release gravity-type vertical engaging bolts. When the lock mechanism itself is fouled up, the best way to proceed is to drill out the cylinder and replace the entire assembly.

Desks. Desks with locking drawers controlled from the center drawer can be opened in a couple of ways besides picking and drilling.

Look at the desk from underneath to see what the locking mechanisms for the various drawers look like. Notice that the locking bar engages the desk by an upward or downward pressure depending on the bolt style. Closing the desk drawer all the way pushes the bolt into the locked position. Herein lies the weakness in the desk's security. The bolt usually needs to be pushed up from under the desk by hand to open most of the drawers. The center one has its own lock.

In other desks, you might have to use a little force and pull outward on the center drawer to push the bolting mechanism downward slightly to open the various drawers.

To open the center lock, follow this procedure:

1. Use two screwdrivers and some tape or cardboard. Put the cardboard or tape between the drawer and the underside of the desk top so you don't mar the desk.

2. Insert a screwdriver and pry the drawer away from the desk top.

3. With the other screwdriver pull the drawer outward to open it. With practice, this can be done with only one screwdriver.

If you drill a small hole in the drawer near the lock, you can insert a piece of stiff wire, such as a paper clip, to push down the plug retainer ring. In doing so, you pull the plug free of the lock causing the bolt to drop down into the open position.

Doors

Possibly the simplest way to open most doors is with a pry bar and a linoleum knife. Insert the linoleum knife between the door and the jamb with the point tipped upward. Insert the pry bar as shown in Figure 17.13. Exerting a downward motion on the pry bar spreads the door slightly and allows you to disengage the locking safety latch.

When this is done, bring the linoleum knife forward, pushing the latchbolt into the locking assembly and opening the door. If there is no safety catch, the knife alone can be used to move the bolt inward. It's also possible to use a standard shove knife or even a kitchen knife.

Sometimes the deadlatch plunger is in the lock but there isn't room to insert a pry bar. What do you do? Use wooden wedges. Insert one of them on each side of the bolt, about 4 to 6 inches from the bolt assembly. Spread the door away from the jamb. Then use a linoleum knife to work the bolt back.

Some doors and frames have such close clearances that you cannot insert a wooden wedge or pry bar. Instead, use a stainless steel door shim. Force it into the very narrow crevice between the door and the frame and work back the bolt.

Often a door lock can be opened with a Z-wire. This tool is made from a wire at least 0.062 inch thick and 10 to 12 inches long (Figure 17.14). Insert the Z-wire be-

Figure 17.13 Opening a door using a pry bar and linoleum knife. (*Desert Publications*)

Figure 17.14 Opening a door with a Z-wire. (*Desert Publications*)

tween the door and the jamb. When the short end is all the way in, rotate it toward you at the top. As you do this, the opposite end will rotate between the door and the jamb. It contacts the bolt and retracts it. If the bolt binds, exert pressure on the knob to force the door in the direction required.

Sometimes, you may be required to open locked chain latches. You can, of course, force the door and break the chain, but there are better ways. The rubber band technique works most of the time.

1. Reach inside and stick a tack in the door behind the chain assembly (Figure 17.15).

2. Attach one end of a rubber band to the tack; attach the other end to the end of the chain.

3. Close the door. The rubber band will pull the chain back. If it doesn't pull the chain off the slide, shake the door a little.

4. If the door surface will not receive a tack, use a bent coathanger to stretch the rubber band (Figure 17.16). Make sure the coathanger is long enough and bent properly so you can close the door as far as possible.

If the door and jamb are even and there is enough space, a thin wire can be inserted to move the chain back.

Sometimes you can open a door very easily if it has a transom. Use two long pieces of string and a strip of rubber inner tubing. The tubing should be 8 to 10 inches long. Attach a string to each end of the tubing so you can manipulate the tubing from the open transom. Lower the tubing and wrap it around the knob. Pull up firmly on both strings to maintain tension and turn the knob. This method can be used with either a regular doorknob or an auxiliary latch unit.

Automobiles

Automobile EEP can be used on front and rear ventilation windows, door windows, doors, and trunks. Automobile locks vary considerably in their design and reliability. Despite this, you can use several methods to gain access. Most of them are faster and easier than picking the lock. In some automobile locks, picking must be ruled out entirely because the sidebar cylinder makes picking a time-consuming, tedious job.

Keyless entry into an automobile is most commonly done at the windows, the most vulnerable part of a car's security. The door release pushbutton lever is often the easiest area to attack, especially on the newer car models that have a single pane of glass. The procedure is fairly simple.

Figure 17.15 The rubber band technique is usually the easiest way to unlock most chain locks.

Figure 17.16 A variation of the rubber band technique.

1. Use a straightened coathanger with a loop or triangle in its end, as shown in Figure 17.17.

2. If the window is rolled up tight, force a paint scraper between the edge of the window and the weatherstripping (Figure 17.18).

3. Insert the coathanger in the crevice made by the paint scraper and pull the pushbutton lever up with the loop or triangle. Of course, a coathanger isn't the only tool that could work here. Any instrument that could slip through the crevice and lift the lever could do the job.

Two 0.22-caliber gun cleaning sets can help you lift either the pushbutton lever or the door handle (Figure 17.19). The cleaning tools come in sections so you can make a variety of rod lengths.

1. Use two gun-cleaning sets to make a rod long enough to reach across the inside of the car.

2. Use the slotted cleaning attachment and some fishing line to make a loop in the end of the rod (Figure 17.20).

3. Work on the window or door opposite the side you are on. Raise the lever or the handle with the loop by pulling on the line and lifting the rod up.

Figure 17.17 Bent coat hangers can be used to lift automobile vertical button door locks.

Figure 17.18 Sometimes a paint scraper can be used to make room for a bent piece of wire or other car opening tool. (*Desert Publications*)

Figure 17.19 A .22 caliber gun cleaning set. (*Desert Publications*)

FISHING LINE

FILE SHARP EDGES SMOOTH

Figure 17.20 A long rod with a closing loop on the end makes a good car lock opening tool. (*Desert Publications*)

The rear ventilation, or wing, window comes in two varieties: with and without a locking button on the swivel. Without the locking button, gaining entry is only a matter of forcing the swivel lock up into the unlocked position.

To open rear ventilation windows without locking buttons on the swivel, insert a putty knife or paint scraper between the window and the frame, bending the knife slightly to allow the unlocking tool to enter. In this case, the unlocking tool could be a thin piece of wire.

Loop the wire around the swivel latch and pull up. The wire should be bent slightly on the end to ensure a better upward motion without it slipping off the lever.

Most cars now have locking pushbuttons in addition to the swivel levers. To unlock these mechanisms you need standard automobile opening tools. Insert these tools on different sides of the lock (Figure 17.21). Use one to depress the button by pulling the tool toward you. Twist the other slightly to push the lever into the unlocked position.

The front ventilation window gives you access to the window roller handle and the door handle. Two different tools can be used depending on the amount of space you have to work (Figure 17.22). Both tools work something like the one shown in Figure 17.23. The window must be pried slightly to insert the tool.

A Lenco car opener tool can also be used from the front ventilation window. You can get the tool through your local locksmithing supply house. The tool can pay for itself with the first lockout that you are called on to open.

How you gain entry into the trunk of a car depends on the lock's location. Drilling is an acceptable practice here. Remember that door locks and the ignition usually take the same key; the glove compartment and the trunk usually take the same key

WINDOW

Figure 17.21 Standard automobile opening tools are used to open rear ventilation windows with locking buttons. (*Desert Publications*)

Figure 17.22 Tools for opening the front ventilation windows. (*Desert Publications*)

Figure 17.23 One method of opening the front ventilation window. (*Desert Publications*)

also. Sometimes you can open the trunk lock by inserting the door key and pulling it out as you turn it.

Picking the trunk lock is simple, but it can be time consuming if you have not worked with many automobile locks. You can knock the lock out by forcing it inward with a hammer and a heavy screwdriver, but this ruins the lock and, sometimes, associated parts on the inside.

Drilling also has drawbacks. You can drill out the plug by shearing off the pins. This means replacing the lock unit. Lately, manufacturers have improved the security of trunk locks by installing a steel plate in front of the catch to prevent drilling. However, it is possible to drill about 2 inches to the side of the catch bolt. This enables you to use a bent piece of wire to force the catch backwards to open the trunk.

There is a retainer in some trunk cylinders. An L-shaped wire inserted into the keyway will catch the retainer, and you can pull it down and pull the plug out of the lock (Figure 17.24). After that a small screwdriver or an awl can be used to push back the catch mechanism.

The easiest of all automobile locks to open is the glove compartment lock. Insert an L-shaped wire into the keyway and pull down the retainer. This wire actually works on two distinct types of locks. In one, you remove the plug; in the other, the wire engages the glove compartment catch mechanism and opens it with a downward movement.

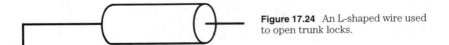

Figure 17.24 An L-shaped wire used to open trunk locks.

Forced Entry

Forced entry in place of proper professional techniques is never recommended except in emergencies or when authorization is given by the owner. There are some simple rules about forced entry you should consider:

- Attempt a forced entry only as a last resort. Try other techniques first.
- If you must jimmy the door, do it carefully. Today, antique doors and frames can be a valuable part of a home. Don't risk a costly lesson by being too hasty.
- If you must break a window to gain access to a lock, break a small one. Replacing broken windows can be expensive.
- Don't saw the locking bolt. This is very unprofessional.

Remember that lock picks and many other entry tools are considered burglar tools in some jurisdictions. In many cases, you must have a license to carry them. Check with your local police, keep your locksmith's license current, and destroy worn-out tools.

Be aware of the trust that your city and customers put in you by allowing you to be a locksmith; you possess certain tools and knowledge that others do not have. Don't abuse this privilege.

18

Combination Locks

Like other locks, combination locks are available in a variety of types and styles. In this chapter we will consider a number of representative types. Some of these locks are more secure than others, but each has its role. For example, it would be inappropriate to purchase a high-quality combination padlock for a woodshed; by the same token, it would be foolish to protect family heirlooms with a cheap lock. Even the best lock is only as good as the total security of the system. A thief always looks for the easiest entry point.

Parts

Though combination locks have some internal differences, all operate on the same principle—rotation of the combination dial rotates the internal wheel pack. The pack consists of three (sometimes four) wheels. Each wheel is programmed to align its gate with the bolt-release mechanism after so many degrees of rotation. Programming may be determined at the factory or it may be subject to change in the field. As the wheels are rotated in order (normally three turns, then reverse direction for two turns, then reverse direction again for one turn), the gates are aligned by stops, one for each wheel and one on the wheel-pack mounting plate. When all the gates are aligned, the bolt is free to release. Padlocks are built so the shackle disengages automatically or manually by pulling down on the lock body. The captive side of the shackle, the part known as the *heel*, has a stop so the shackle will not come free of the lock body.

One weakness of low-priced locks is their ease of manipulation. With practice you can discriminate between the clicks as the wheels rotate. Once you learn to do this, manipulation of the lock is child's play. Fortunately, manufacturers of better locks make false gates in the wheels that make manipulation much more difficult. A real expert will not be thwarted, but experts in this arcane art are few and far between.

It takes training and intensive practice to distinguish between three or more false gates on each wheel and the true gate.

Manipulation

Manipulation of combination locks is a high skill—almost an art. A deep understanding of combination-lock mechanisms and many, many hours of practice are needed to master this skill. But the rewards—both in terms of the business this skill will bring to your shop and the personal satisfaction gained—are great.

No book can teach you to manipulate a lock any more than a book can teach you to swim, box, or do anything else that is essentially an exercise in manual dexterity. But a book can teach the rudiments—the skeleton, as it were, that you can flesh out by practice and personal instruction from an expert.

Manipulation is a matter of touch and hearing. It also involves an understanding of how the lock works. Electronic amplification devices are available to assist the locksmith. These devices are useful, especially as a training aid.

Padlocks usually do not have false gates and so are ideal for beginners (Figure 18.1). Pull out the shackle as you rotate the wheels. This tends to give better definition to the clicks.

Work only for one number at a time. Stop when the bolt hesitates and touches the edge of the gate. The bolt has touched the far side of the gate, so move back a number on the dial and note it. Turn the dial in the reverse direction, going past the first number at least twice. Pull on the shackle and slowly continue to turn the dial. When you sense that the bolt has touched the gate, note the number that came up just before the bolt responded. This is the second number of the combination. The third number comes easier than either the first or second one. The dial may stop at either

Figure 18.1 A typical dial combination padlock in exploded view. (*Desert Publications*)

the first or second number. Since many of the earlier combination locks did not have the accuracy of modern locks, the bolt catches at any one of the three numbers at any time.

You have the three numbers, but perhaps not in their correct order. Vary the sequence of the numbers until you hit the right one.

Because of imperfections in older and inexpensive modern locks, the bolt may stop at points other than the gates. Only through practice can you learn to distinguish between true and phantom gates.

For practice, obtain two or three locks of the same model. Disassemble one to observe the wheel, gate, and bolt relationship and response. These insights will help you manipulate the other two locks.

Drilling

As a last resort, all combination locks can be opened by drilling. To drill these locks follow this procedure:

1. Drill two ⅛-inch holes in the back of the lock.
2. Turn the dial and determine that the gate of one wheel is aligned with one of the holes. You might be able to see the gate. If that fails, locate the gate with a piece of piano wire inserted through one of the holes.
3. When the gate and the hole are aligned, note the number on the dial face. Determine the distance, as expressed in divisions on the dial, between the hole and bolt.
4. Subtract this distance from the reading when the gate is aligned with the hole. The result is the combination number for that wheel.
5. Reverse dial rotation and find the number for the second wheel. Do the same for the third.

Changing Combinations

Many locks are designed for combination changes in the field. Three of them are described here.

Sargent and Greenleaf

S & G padlock combinations are changed by key.

1. Turn the numbers of the original combination to the change-key mark located 10 digits to the left of the zero mark on the dial face.
2. Raise the knob on the back of the lock to reveal the keyway. Insert the key and turn 90°.
3. Repeat step 1, but use the new combination and the change-key mark as zero.
4. Remove the change key and test the new combination.

Simplex

The Simplex is a unique combination lock, employing a vertical row of pushbuttons rather than the more usual dial (Figure 18.2).

1. Turn the control knob left to activate the buttons.
2. Release the knob and push the existing combination (Figure 18.3).
3. Push down the combination change slide on the back of the lock (Figure 18.4).
4. Turn the control knob left to clear the existing combination (Figure 18.5).
5. Push the buttons for the new combination firmly and in sequence (Figure 18.6).
6. Turn the control arm right to set the new sequence (Figure 18.7).

Dialoc

Partial disassembly is required to change the combination. Refer to Figure 18.8 for the parts numbers in the following steps.

Figure 18.2 The Simplex pushbutton combination lock. (*Simplex Access Controls Corp.*)

Figure 18.3 Changing Simplex combination, Step 2.

Figure 18.4 Changing Simplex combination, Step 3.

Figure 18.5 Changing Simplex combination, Step 4.

Figure 18.6 Changing Simplex combination, Step 5.

Figure 18.7 Changing Simplex combination, Step 6.

Figure 18.8 The very sophisticated Dialoc 1400. (*Dialoc Corporation of America*)

To disassemble:

1. Remove the Dialoc from the door by withdrawing the two mounting screws 487-1, holding the inside plate 821 against the door.

2. Place the outside plate 820 face down on a smooth work surface so that housing 414 is facing up.

3. Remove the three nuts 484-5 holding the housing cover 745.

4. After removing the nuts, gently lift the housing cover up and away from the housing. Take care that control-bar spring, 744 is not dislodged and lost in the process. Lay it where it won't get lost.

5. Remove the secondary arm 402.

6. Grasp the end of the control bar 726 and lift out the four ratchet assemblies. Observe how they are arranged before dismantling.

Now you can rearrange these four ratchet assemblies to get a new combination. (Should you want to use numbers that are not in the present combination, your distributor or the Dialoc Corporation can readily supply these.) Remove the ratchet assemblies 469 from the control bar and change the sequence as you desire. Put the ratchet assemblies back on the control bar. When replacing the ratchet assemblies on the control bar, make sure they go on with the ratchet part of each assembly down and that the first digit of the combination is put on first, then the remainder of the combination in order from the first ratchet assembly outward.

To assemble:

1. Place the four ratchet assemblies 469 on the control bar 726 and, grasping the assembled control bar and the ratchet assemblies, slip them over the center spindle of the primary arm 406. The ratchet assemblies can be guided over the lower mounting pin protruding from the outside plate 820.

2. Take care that the control bar 726 passes through the elongated hole in the bottom of the housing 414.

3. Replace the secondary arm 402 on the primary arm 406. Make sure that the finger 408 and the finger follower spring 412 are in position.

4. If the cam pawl trip 441 was dislodged during the combination change, replace it by inserting it into the housing with the small protruding end down. The blade on this part must be against the ratchet pawls of the ratchet assemblies.

5. Make sure that the secondary arm 402 is flush with the top of the square bar. This allows a spacing of approximately 0.020 inch between the secondary arm and the top ratchet assembly.

6. After the secondary arm is properly placed, return control spring 744 to the end of the control bar 726.

7. Replace the housing cover 745. Make sure that the cam pawl trip 441 and the control bar 726 are in their respective holes in the housing cover. A straight pin is

helpful in guiding these parts into holes. Be sure reset arm 870 is inside the housing 414.

8. Replace the cover-retaining nuts 484-5. Turn these nuts snugly with your fingers, plus approximately one-half turn with a wrench.

9. Operate the lock before you replace the door to ensure its proper operation on the new code settings.

Master 1500 padlock

The Master 1500 combination padlock is covered with heavy-gauge sheet steel, rolled and pressed at the edges. The net result of this somewhat unorthodox construction is a tight-fitting lock and one that is not easily pried open. Disassembly is recommended only as a training exercise. It is not practical to open this lock for repair.

Three wheels are employed, each with a factory-determined number. If you receive the lock with the shackle open you can determine the combination by looking through the shackle hole. Note the dial reading as you align each wheel. If you read the numbers properly and if the wheel gates are in alignment, you should only have to add 11 to the dial readings to get the true combination. If you are slightly off, compensate by adding 10 or 12 to the original readings.

The Master 1500 series can be masterkeyed (Figure 18.9). This lock is popular in schools. While the combination can be obtained by manipulation, picking the keyway is faster. Once the lock is open, determine the combination by the method described above.

Other Keyless Locks

Combination locks are those devices that do not require a key to open. Although many people think of such things as safes and vaults or perhaps a bicycle padlock requiring a three-digit combination as the only types of combination locks, this is not true. Any device that does not require a key but instead relies on a single or series of digits or letters in a certain order to open the mechanism, is classified as a combination locking device.

Figure 18.9 The Master 1525 masterkeyed dial combination padlock. (*Master Lock Company*)

The following sections discuss, illustrate, and provide specific details concerning several of the best and most popular keyless locking devices available today. Remember that the products discussed here are not the only types, nor are they representative of the only manufacturers of such products; they are the products that have proven themselves to a wide variety of customers in varied and numerous types of installations. As such, they are recommended for your consideration as some of the highest quality, professional, and most secure keyless combination locking devices on the market today.

Simplex keyless locks

Keyless entry locks provide exceptional security in a variety of modes. Motels, hotels, various private and government institutions are using these locks with increasing success (Figure 18.10). The main manufacturer and proponent of keyless locking devices is Simplex Security systems. Their locks are commonly used in a variety of security-related situations where more than one entrance control device is required. As an example, consider the case of a company research and development facility. For added security, a keyless lock is added to each door. At night, then, two locks secure the door. During the day, the main lock is in the unlocked position, but the pushbutton lock is readily available for the use of the research and development employees because they are the only individuals who know the particular combination to gain access.

In motels and hotels, the use of keyless locks has increased security, cut costs, and saved time and manpower. How? By using the Simplex lock, in one example, one motel stopped losing more than 50 keys a month and saved the costs of continually having duplicate keys made and locks rekeyed. Best of all, thefts were almost entirely eliminated.

Figure 18.10 Simplex 5 pushbutton deadbolt lock. (*Simplex Access Controls Corp.*)

Time and manpower were also saved with the installation of the Simplex lock. The desk clerk no longer had to hunt a duplicate key or go to a room and unlock it because the key was locked inside. The clerk simply gives the specific room combination, just as he did when the guest checked into the motel. Once the guest checks out, it requires only three minutes to change the combination; this keeps room security integrity very high.

The Simplex lock is extremely easy to operate and maintain, despite its very high security. The lock has five pushbuttons arranged in a circle and numbered from one to five. The combinations can be set at random; to operate the lock, the various buttons are pushed in sequence, in unison, or a combination of both.

Installation of the lock is also relatively easy. The only tools required are a screwdriver, an electric drill, and a hole-saw set.

Advantages. Before we get into the various installation procedures, let's look at the specific advantage of a Simplex lock so you will understand, as a locksmith, why many businesses and homeowners are installing these locks.

- No keys are required. For businesses, this means none are issued or controlled and there are none to recover when the employee leaves.
- They are burglar-proof because there is no keyhole to pick.
- There are thousands of possible combinations.
- You can change the combination in seconds whenever there is a change in personnel or tenants or for other reasons of security.
- Money is saved because there are no keys or cylinders to be changed or replaced.
- The locking units are extremely durable. They are wear-tested for the equivalent of 30 years of intensive use.
- It is an extremely easy-to-install lock, more so than any other existing lock or knob type lockset. It is completely mechanical and requires no electrical wiring or specialized accompanying equipment.

Varieties. The pushbutton combination lock (Figure 18.11) comes in several models: the deadlock with a full 1-inch deadbolt, the automatic spring latch, the key bypass, and the special security model (for use with Department of Defense or industrial complexes requiring the security necessary for the protection of closed or restricted areas).

The *deadlock*, model DL, requires only the combination pressed in the right order and the turn knob moved to open the lock from the outside. The lock remains in the open position, with the bolt retracted, until either the inside lever or outside knob is turned to throw the bolt into the lock position.

The *automatic spring latch*, model NL, locks automatically. It opens with a finger lever from the inside. A thumb slide on the inside holds the retracted latchbolt in the unlocked position, if desired (Figure 18.12). A variation, model NL-A, lacks the latch holdback. A deadlocking latch is standard for these units; this prevents prying the latch back on the closed door. The latchbolt is reversible without affecting the deadlatching function.

Figure 18.11 Deadbolt with a key bypass. (*Simplex Access Controls Corp.*)

Figure 18.12 Finger lever and thumb slide for the Simplex Modl NL lock; inside view of door and back of lock. (*Simplex Access Controls Corp.*)

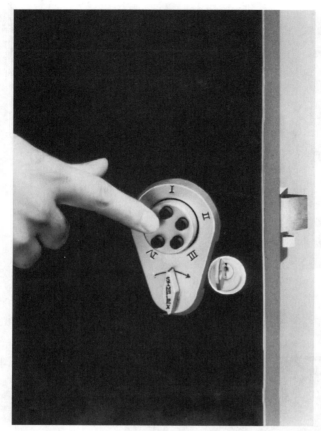

Figure 18.13 The key bypass is also available with the standard latch bolt model Simplex lock. (*Simplex Access Controls Corp.*)

The *key bypass* models (Figure 18.13), DL-M and NL-M, are the DL and NL models with an optional key cylinder. They provide the flexibility of a master key system and the security advantages of a combination lock. Employees, tenants, and others use the combination. The key ensures entry regardless of the combination setting used, but is given only to designated management personnel. This eliminates the need to maintain a central listing of combinations for reference use by top management personnel. It is completely safe because there is no widespread proliferation of keys and key distribution can be severely restricted.

The key bypass model is popular in facilities working for and with the Department of Defense and other government agencies, due to its excellent security advantages.

The special security model, NL-A-200-S, is designed for complete compliance with the Industrial Security Manual of the Department of Defense. The model is an approved substitute for the expensive guard forces in the protection of closed and restricted areas in contractor facilities. It has all the features of the model NL automatic spring latch, plus

- A face plate shield to prevent observations of pushbutton operation and maintain the secrecy of the combination (Figure 18.14).

- The combination change access can be padlocked to ensure authorized combination changes only.

- The latch holdback feature is eliminated so the lock can never be kept open. It is in locked condition at all times.

Installation.   All of the Simplex models have the same basic three-piece self-aligning assembly and install easily and quickly. *Note:* All Simplex door locks require a minimum flat surface of 3½ inches on the inside of the door to mount.

Simplified, the procedures for installation are as follows.

Figure 18.14 Face plate shields prevent observation of the combination code by unauthorized individuals. (*Simplex Access Controls Corp.*)

Figure 18.15 Standard and optional strikes. (*Simplex Access Controls Corp.*)

Wood and door frame installation. Drill two holes, ¾ inch and 1⅝ inches in diameter, a distance of 2⅝ inches from the door edge. Key bypass models will require an additional ⅞-inch key cylinder hole. All parts are provided for a complete wood door and frame installation. Included are two styles of strikes (standard) to suit any wood door frame (A in Figure 18.15).

Metal door and frame installation. An adapter kit containing a special surface-mounted strike (B in Figure 18.15) is available. The special strike is fastened to the inside surface of the metal frame and eliminates the need to cut or mortise the frame for the standard strike. The adjustable riser plate raises the lock housing to ensure the proper alignment of the latch or deadbolt with the special strike.

Models DL and NL installation

1. Using the template provided with the lock (Figure 18.16), cut the template from the sheet and fold on the printed line as shown by the arrow. With the door closed, tape the template to the inside of the door with the fold line aligned with the edge of the door (or door stop if one is present). The same template applies to both inswing and outswing doors (Figure 18.17). Using a center punch, mark the centers of the two large and three small holes.

2. Using a ⅛-inch-diameter drill, drill the centers of the ¾-inch-diameter and 1⅝-inch-diameter holes all the way through to the front surface of the door. Drill the three mounting screw holes approximately ½ inch deep.

3. With a 1⅝-inch-diameter hole saw, open up the larger hole by drilling from both the outside and the inside surfaces, meeting in the middle. Follow the same pro-

cedure with a ¾-inch-diameter drill for the smaller hole. *Note:* With Model NL, the latch is reversible by removing the two large and two small screws on the undersurface and disassembling the lock housing.

4. Slide the lock housing into the holes in the door as shown in Figure 18.18. If the holes are drilled at an angle and the undersurface of the lock is not in full contact with the door, file out the holes to correct. Screw the lock housing to the door.
5. Place the holding bracket on the front surface of the door with the slotted legs engaging the aligning pin on the barrel. With the large radius of the holding bracket mating with the radius of the barrel, fasten with the screws provided.
6. Remove the lock housing from the door.
7. Slide the holding edges of the face plate behind the formed up edges of the holding bracket. Make sure the face plate is securely held on both sides.
8. Replace the lock housing while holding the control knob, marked "SIMPLEX," in the vertical position. Turn the knob to ensure proper engagement and fasten the lock housing to the door.

Note: It is very important to try the combination of the unit several times before locking. It is also necessary to have your customer open the lock several times before you leave. If the lock does not fit exactly to the door, there is an adjustable riser plate

Figure 18.16 Drilling template for the 100 and 200 series DL/NL model locks. (*Simplex Access Controls Corp.*)

Fold template on **Dash Line** for additional clearance required on outswing doors or when using Part #100 Riser Plate on inswing doors.

Fold template on **Solid Line** on inswing doors when **not** using Riser Plate.

OUTSWING

INSWING

Figure 18.17 This visual provides for the proper determination of which template line will be used for positioning the lock properly to the door. (*Simplex Access Controls Corp.*)

For all NL Latch type locks clearance between lock housing and strike must not exceed ⅛" to ensure deadlatching.

Figure 18.18 Lock housing and holding bracket positioning for proper lock installation. (*Simplex Access Controls Corp.*)

available from the factory. With minimum maintenance, this unit will now provide a very long and extremely useful service to your customer.

Key bypass installation instructions. Follow the above instructions for installing the standard DL and NL locks, with these exceptions:

- Use the template (Figure 18.19) for positioning a ⅝-inch-diameter hole for the key lock cylinder.

- With two pairs of pliers, break the tailpiece of the key cylinder at the proper location as indicated (Figure 18.20): (1) for the Simplex 100 or 200 series locks when the riser plate is not used; (2) for 100 series when either ¼-inch or ⅜-inch side of the riser plate is used, or with the 200 series locks when ¼-inch side of the riser plate is used; (3) for 200 series locks with ⅜-inch side of riser plate, use full length of tailpiece.

- Assemble the key lock cylinder in the ⅞-inch-diameter hole. *Note:* The hole in the flange of the cup must be positioned over the ⅛-inch-diameter mounting screw hole used to secure the lock housing to the door (Figure 18.21).
- Proceed with the last step of the basic instructions.

You may wish to substitute a standard rim cylinder for the ⅞-inch-diameter cylinder supplied with the Simplex key bypass models DL-M and NL-M. Due to space limitations, the Simplex face plate with Roman numerals must cover a portion of the standard key cylinder (Figure 18.22).

Figure 18.19 Drilling template for the key bypass feature. (*Simplex Access Controls Corp.*)

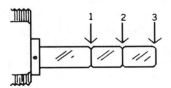

Figure 18.20 Tailpiece break points. (*Simplex Access Controls Corp.*)

Figure 18.21 Bypass lock assembly procedure. (*Simplex Access Controls Corp.*)

Figure 18.22 Simplex lock using a standard sized rim cylinder. Note how the Simplex overlaps a portion of the cylinder. (*Simplex Access Controls Corp.*)

When a standard key cylinder is used, you must let the factory know this; in your order you must specify the Simplex model (DL-M or NL-M, 200 series, and whatever finish is desired) plus the very important notation ". . . for use with standard rim cylinder . . ." placed prominently on your order blank.

Make sure you get the *rim cylinder*, not the *mortise*. In the proper keying of the unit, your local or regional distributor or other supply source will probably be contacted also, unless you have a wide variety of various lock parts in stock. You must furnish the cylinder. An escutcheon (or bezel) must be used that allows the nose of the cylinder to be mounted flush with the front surface of the door (Figure 18.23). Cut the length of the tailpiece to suit. Cut or grind the width of the tailpiece to $\frac{3}{16}$ inch to fit the tubular coupler on the Simplex lock.

Use the Simplex key bypass template (Simplex Form 2-15-65A) for drilling. Use the center of the $\frac{7}{8}$-inch-diameter hole for drilling the appropriate size hole for the standard rim cylinder and escutcheon. Assemble the cylinder, escutcheon and back plate to the door *before* mounting the Simplex lock.

Install the Simplex by following the standard installation instruction procedures. The Simplex face plate (with Roman numerals) and holding bracket will cover part of the cylinder. It might be necessary to notch the edge of the escutcheon to clear one of the two screws used to fasten the holding bracket to the front surface of the door. One of the two lock housing screws nearest the door edge must be fastened to the back plate of the rim cylinder (Figure 18.24). Using the hole in the lock housing as a guide, drill a 0.161-inch-diameter hole (#20 size drill) in the back plate for the #10 sheet metal screw.

Servicing Simplex locks. After you install the Simplex lock, you might have a service and maintenance contract for the numerous Simplex locks in the building. During the course of your locksmithing career, you will be called on many times to carry out your portions of the service contract with the building manager. Among the several procedures most common for the Simplex lock, you may change the combination for one or many of the locks or disassemble the lock for one reason or another. You will also be expected to find an unknown combination. The procedures for these various parts of the locksmith's job are detailed below.

Combination changing. With the door in the open position and the lock in the locked position, push the existing combination. Remove the hexagon screw at the top of the lock housing (Figure 18.25) with an allen wrench. After removing the screws, insert the wrench into the hole and depress the red button inside. Remove the wrench.

Figure 18.23 Standard lock cylinder positions with the door front surface. (*Simplex Access Controls Corp.*)

Figure 18.24 Arrow indicates lock housing screw that must be attached to the cylinder back plate. (*Simplex Access Controls Corp.*)

Figure 18.25 Use the Allen wrench to remove the access screw to change the combination of the lock. (*Simplex Access Controls Corp.*)

Figure 18.26 Simplex Unican 1000 Series Heavy Duty Lockset. (*Simplex Access Controls Corp.*)

Turn the front control knob (below the pushbuttons and marked Simplex) to the left. This removes the existing combination.

Depress the pushbuttons on the front in the sequence desired for the new combination, (This can be up to five single numbers or two at a time simultaneously.)

Write down the combination sequence immediately and present it to your customer for safekeeping. Turn the front knob (marked Simplex) to the right. The new combination is now set in the lock. Try the combination several times with the door in the open position before attempting it with the door in the closed position and locked. (Stand inside the door and let the customer operate the combination at least once from the outside.)

The Simplex Unican 1000 series lock

The Simplex Unican 1000 series lock (Figure 18.26) is a heavy-duty lockset ideal for securing high traffic areas in commercial, institutional, and industrial buildings, apartment lobbies, hotels, motels, luxury homes, and restricted areas in commerce and industry. Like other high quality Simplex products, the Unican 1000 has the following features:

- Built-in security with thousands of possible combinations
- No keyway to pick
- Weatherproofing which allows for all types of installations
- A ¾-inch deadlocking latch with a 2¾-inch backset
- Completely mechanical

- No electrical wiring
- One-hand operation

The unit is 7¾ inches high, 3 inches wide, and 3¹³⁄₁₆ inches deep. Latch front is 1⅛ inches × 2¼ inches with a backset of 2¾ inches. The Unican lock will fit doors with thicknesses ranging from 1⅜ inches to 2¼ inches.

A key bypass option is available, mainly for masterkeying purposes, using the Best or Falcon removable core cylinders in the outside knob (Figure 18.27). The removable core cylinders are not available with the lock when ordered; they must be obtained separately. The Russwin/Corbin removable core capability can be obtained on a special order basis.

The second optional feature is the passage set function. When activated by an inside thumb turn or key cylinder, the outside knob can be set to open the lock without the need for using the pushbutton combination (Figure 18.28). This is designed for use in offices, etc., when the door is to be left unlocked for an extended period of time.

Figure 18.27 Unican 1000 with key bypass fester. (*Simplex Access Controls Corp.*)

Figure 18.28 Key bypass or thumb turn can be employed with the unit. (*Simplex Access Controls Corp.*)

Both 9-inch and 14-inch high "mag" type filler plates are available to cover up previous lock installation holes on 1¾-inch doors. These three-sided steel plates wrap around the door extending 5½ inches on both the outside and inside surfaces.

Installation. The Unican 1000 uses an ASA 161 cutout (2⅛ inches in diameter through the hole with a 2¾-inch backset) like most other heavy-duty cylindrical locksets (Figure 18.29). In addition, the lock requires two ¼-inch-diameter holes to bolt the top of the lock case to the back side of the door and a 1-inch-diameter hole for the key-operated combination change mechanism to operate from the back side as well. The Unican can be installed in wood or metal doors from 1⅜ inches to 2¼ inches thick.

In addition to the special tool provided with each lock, you, the installer, also need the following tools:

- Drill (½-inch-diameter electrical drill preferred)
- 2⅛-inch-diameter hole saw (for metal doors) or spur bit (for metal doors)
- 1-inch-diameter hole saw or spur bit
- ¼-inch-diameter drill
- Two wood chisels (wood doors), ¼ inch and 1 inch wide
- Phillips and standard screwdrivers
- Center punch and hammer
- Standard 12-inch ruler

The template. Before starting the actual installation procedures, let's look at a portion of the Unican 1000 three-part template that will be used. (Figure 18.30 shows the template and proper dimensions; Figure 18.31 shows the door edge portion of

Figure 18.29 Exploded view of Unican 1000 positioning for a wood door. (*Simplex Access Controls Corp.*)

Figure 18.30 Unican 1000 door template. (*Simplex Access Controls Corp.*)

Figure 18.31 Unican 1000 template portion for marking a door thickness. (*Simplex Access Controls Corp.*)

the template.) Figure 18.32 provides the information and specifications for the latch cutout required. *Caution:* When using this type of template, always apply the template and drill from the outside, being sure to compensate for the door bevel, if any (Figure 18.33).

Figure 18.32 Unican 1000 latch cutout specifications. (*Simplex Access Controls Corp.*)

Figure 18.33 Details of the bevel compensation for the Unican 1000. (*Simplex Access Controls Corp.*)

Basic installation. Initially, make all necessary corrections to ensure the door is properly hung in the frame. Apply the template to the high edge of the door bevel and tape it in place. If a metal door frame pre-positions the strike location, position the template to center the 1-inch-diameter latchbolt hole in the strike cutout in the frame. With a center punch and hammer, mark the centers of all required holes.

With the ¼-inch drill, drill the centers of the 2⅛- and 1-inch holes all the way through the door. If you want to center mark the hole on the other side of the door (the low side of the bevel) using the template, be sure to move the "fold" line to shorten the backset to compensate for the bevel. Lay a carpenter's square on the high edge of the bevel to see how much compensation to make.

It is a good practice to bore the 2⅛- and 1-inch diameter holes halfway in from both the front and back surfaces of the door.

If the door was originally prepared for a 2⅜-inch backset, file out the 2⅛-inch hole by an additional ⅜ inches; the lock case and inside rose are wide enough to cover.

Be sure to drill the 1-inch latchbolt hole on the door edge straight and level to prevent binding of the latch assembly. Carefully drill the two ¼-inch mounting holes through the door. *Note:* A drill jig is of great assistance in keeping your mounting hole straight and true.

Take care to allow clearance for the hinge that operates the antifriction device on the Unican latch. The latchbolt hole must be relieved adjacent to the antifriction trigger to permit the hinge to rotate freely and prevent the binding of the latch.

For a proper Unican 1000 installation, it is imperative to have a freely operating latch properly positioned to the strike. Excessive binding or dragging of the latch can cause the front knob clutch to *freewheel*, preventing latch withdrawal. Also, the deadlatching plunger must be held depressed by the strike plate to be operable. The gap between the door and the frame must never be more than a ¼ inch to ensure proper deadlatching. If door silencers are present, allowances must be made in positioning the strike.

Position the strike. To properly position the strike, install the latch assembly into the edge of the door. On wooden doors, mark around the face plate of the latch and mortise in with a chisel. Remove ⁵⁄₁₆-inch minimum for the antifriction trigger hinge and ⁵⁄₃₂ inch for the face plate assembly. When inserted, the latch assembly should be flush with the nose of the door. Close the door to the point where the latch touches the door frame and mark the frame at the top and bottom lines previously scribed. Mark the screw holes. When the door is closed, the screw holes of the strike should be even with the screw holes of the latch plate. Cut or mortise the frame to a minimum depth of ¾ inch to ensure a full throw of the latch when the door is closed. Combinations cannot be cleared or changed if the full travel of the latch is impaired. The Unican 1000 comes factory packed with a 2¾-inch strike (and strike box for metal frames). An ASA 4⅞-inch strike is also available as an accessory from the manufacturer. Do not use a strike with a lip radius different than that supplied by the manufacturer. A short radius on the lip can damage the latch assembly.

Changing the hand. Unless otherwise specified in your order, all Unican 1000 series locks are factory assembled for left-hand operation. To change it to right-hand operation, remove the back plate containing the cylindrical unit from the lock case by removing the Phillips head screws. Remove the four Phillips head screws holding the cylindrical unit at the back plate (Figure 18.34) and rotate the cylindrical unit so that the cutout for the latch faces in the opposite direction. Reattach the cylindrical unit using the four screws (with two lock washers for each screw), making sure the unit is centrally positioned to the hole in the back plate and rotates freely when the spindle is turned. To prevent internal jamming, two lockwashers must be used with each screw. Remount the back plate assembly to the lock case.

Adjusting for different door widths. All locks are factory set for a 1¾-inch-thick door. To adjust for a 1⅜-inch to 1½-inch door, remove the back plate and cylindrical unit assembly from the lock case and remove the spacer that is positioned between the cylindrical unit and the back plate (Figure 18.35). Reassemble the cylindrical unit with the shorter (⁵⁄₃₂-×-¼-inch) screws provided with the unit. Insert the roll pin in

Figure 18.34 The four screws must be removed so the cylinder can be rotated, if necessary. (*Simplex Access Controls Corp.*)

Figure 18.35 Backplate assembly and cylinder unit must be removed to adjust the lock for door thickness variations. (*Simplex Access Controls Corp.*)

the front knob shaft one hole closer to the front knob (from position #2 to #3 in Figure 18.35). Again, be sure that there are two lockwashers under each screwhead joining the cylindrical unit to the back plate.

To adjust for 2- to 2¼-inch doors, follow the previous instructions but add the extra spacer provided (lip edge to lip edge) to the one already assembled to the lock. Use the longer (⁶⁄₃₂-×-¾-inch) screws provided, and insert the pin in the shaft in the hole farthest from the front knob (from position #2 to #1) in the shaft in the hole farthest from the front knob (from position #2 to #1 in Figure 18.35). When remounting the cylindrical unit, be sure it is centered in the hole in the back plate and turns freely. Ensure that the two lockwashers are under each screwhead.

Hold the latch assembly in the edge of the door and slide the lock case in from the front. Depress the latch slightly to engage the "T" (A in Figure 18.36) in the lock retracting shoe (B). Viewing the unit from the back side of the 2⅛-inch hole, be sure the curved lips of the latch housing slide within the opening of the cylindrical case (C) in Figure 18.36. Hold the lock case to the front of the door with one hand and insert the two top mounting screws from the back. *Finger-tighten* both screws to ensure the holes were properly positioned. If you are not able to start the screws into the back of the lock case with your fingers, the mounting holes were not positioned correctly. Forcing can strip the threads. To correct this condition, remove the lock case and file out the holes to ensure proper alignment of each to the nuts on the back plate.

Screw on the inside rose and tighten it with the spanner wrench. Slide the knob onto the inside sleeve, depressing the knob catch spring by pushing the pointed tip of the wrench through the hole in the side of the knob collar. When the knob is pushed into the stop, remove the tool and make sure the knob is securely fastened.

Mount the trim plate horizontally over the 1-inch-diameter combination change hole and the two mounting screws (Figure 18.37). With the center punch, indent the door slightly for the two protrusions on the plate used to prevent rotation. Cut the lock screw to length as required. Insert the control key and keep turning clockwise to thread the lock screw into the combination change sleeve assembly on the back plate of the lock case. Tighten the screw until the plate is snug to the door, but do not overtighten. The key can only be removed in a vertical or horizontal position. Overtightening may jam the lock screw in the change sleeve, requiring the removal of the screw with your needle-nosed pliers.

Changing the combination. All locks are factory set for the same combination: buttons #2 and #4 are pushed together, then #3 individually. For security reasons, you should change the combination immediately, using as few as one button or as many

Figure 18.36 Latch assembly with critical portions indicated; these three points (A-B-C) must be worked in exact order to ensure proper lock installation. (*Simplex Access Controls Corp.*)

Figure 18.37 The trim plate must be positioned horizontally over the hole in the door. (*Simplex Access Controls Corp.*)

as five. When the buttons are set for pushing at the same time with other buttons, it is a different combination than if the buttons are set to be pushed individually (i.e., 1, 2, 3 individually pushed buttons is a different combination than if 1, 2, 3 are pushed at the same time). The same button cannot be used more than once in any given combination. If your Unican has a passage set function, be sure it is not in use when the combination code is changed.

To set the combination, the door must be open. Remove the inside trim plate with the control key. Turn the front knob fully clockwise, then release. Push the existing combination, then release the buttons. The buttons should never be held depressed during lock opening. Insert the flat blade end of the special tool into the slots of the combination change sleeve and rotate clockwise to stop, but do not force. Then, rotate counterclockwise to stop and remove the tool. The combination change sleeve will rotate easily when the correct combination is pushed. It will not rotate with an incorrect combination, and forcing can damage the mechanism.

Turn the front knob fully clockwise to stop and release. Turn only once at this stage of the combination change. Depress the buttons firmly and deliberately in the sequence desired for the new combination. Record the combination at once so it will not be forgotten. Turn the front knob fully clockwise to stop to set in the new combination. Try the new combination before closing the door and be sure you have properly recorded it. *Note:* If the Unican will open without pushing a code, you are in the "0" combination mode because the front knob was turned more than once or turned out of sequence. In such a case, repeat all the steps in the change procedure *except* pressing the existing (old) combination. It is very important that the front knob be turned fully *clockwise to stop* when called for in the combination change procedure to ensure that the old code is completely cleared out and the new code is completely set in. Full rotation will be impaired by any of the following:

- Using a latchbolt assembly not supplied by the manufacturer (especially one with a shorter backset or shorter throw)
- A shallow strike box that prevents the full ¾-inch throw of the latch when the door is shut
- Any binding of the latch assembly in the door or in the strike caused by improper door or frame preparation or by door sag and/or buckling

Further installation tips:

- MAG-type filler plates (Figure 18.38) in 9-inch and 14-inch lengths are available for covering up existing mortise lock preparation or badly butchered doors. These can be obtained for 1¾-inch doors only. They are handy to use for an installation template as well.
- Do not allow either the inside or the outside knob to hit against the wall. A floor mounted door stop should be used to prevent damage to the factory adjusted clutch mechanism behind the knob. The manufacturer's one year warranties are voided if the knobs are allowed to hit. If there is a door closer, be sure that it is properly adjusted for the Unican.

Figure 18.38 Filler plate.
(*M.A.G. Engineering and Mfg., Inc.*)

- Due to its unique design, clockwise rotation of the front knob performs two functions. It clears out any incorrect combination attempts and reactivates any buttons previously pushed, and it withdraws the latchbolt if the combination has been pushed properly. When operating the lock, *never* hold the buttons depressed while turning the front knob; damage to the unit can result. Just depress the buttons until they stop and then release them.
- Proper installation and operation of the Unican will result in many years of satisfied customer usage. Just remember that it starts with a good installation.

Servicing. You might be required to service the Unican 1000 series locks. In the majority of instances, servicing is the result of customer mishandling and abuse of the unit. Once in a very great while the unit itself may malfunction for some unknown reason. At this time, you will require some basic information concerning the various parts of the Unican 1000 unit and illustrative material for the proper assembly or disassembly. Figure 18.39 is an exploded view of the Unican 1000. Figure 18.40 is a parts item description and the assembly number and parts.

All Simplex auxiliary locks are shipped with the following combinations: #2 and #4 pushed at the same time, then #3 pushed singly.

Some people forget how to operate this lock, even though the procedure is very simple:

1. Turn the control knob to the left to activate the buttons.
2. Press the correct buttons in the proper order.
3. Release the last button(s) before turning the control knob. Turn the control knob to the right to open.
4. To relock, turn the control knob to the left. Model NL locks automatically.

Some customers will look at the lock they contemplate purchasing and wonder about the control knob. Since it operates the bolt mechanism, can the lock be

Figure 18.39 Exploded view of the Unican 1000 lock unit. (*Simplex Access Controls Corp.*)

opened if it is removed? The answer is a resounding no. The front control knob cannot be forced to open the lock since it is connected to the lock housing by a friction clutch. If the knob has been forced, it will be at an angle and can be turned back to the vertical position by hand or with a pair of pliers without damaging the lock.

Preso-Matic

The Preso-Matic keyless pushbutton combination lock comes in either a deadbolt version (Figure 18.41) or a deadlatch version (Figure 18.42). This is another completely mechanically operated lock; no electricity is required. From the outside, the customer presses the selected four-digit combination. The deadbolt or deadlatch automatically retracts, opening the door. With the deadlatch model, the lock automatically locks each time the door is closed.

Several models are available, so each customer can have a version with the appropriate options to meet his special need. All models come with an instant-exit feature that requires only the depressing of an unlock button on the door inside (Figure 18.43). Also available is a nightlatch button, which provides an added measure of protection for the home or apartment owner or the businessperson working late at night. It means that nobody is going to interrupt (or surprise) him at work or at home.

ITEM NO.	PART NUMBER	DESCRIPTION	PER UNIT
1	200024	Push Buttons	5
2	200026	Bushing,Shaft	1
3	200079	Spring, Nut	1
4	200080	See Assembly Chart	1
5	201006	Link, Combination Chamber	1
6	201016	Spring, Balance	1
7	201018	Spring Double Wave	1
8	201019	Knob Return Spring	1
9	201037	Inside Trim Plate Only	1
10	201038	Pin, Linkage	1
11	201040	Stop Plate	1
12	201048	See Assembly Chart	1
13	201049	Front Plate	1
14	201050	See Assembly Chart	1
15	201093	See Assembly Chart	1
16	201102	Spacer Door Thickness	2
17	201144	Latch	1
18	201155	See Assembly Chart	1
19	201218	Drive Insert	1
20	201267	See Assembly Chart	1
21	201272	Shoe	1
22	201275	Inside Sleeve	1
23	201276	Shoe Retainer and Bridge	1
24	201277	Cover for Shoe Housing	1
25	201278	Installation Wrench	1
26	201280	Rose	1
27	201281	Rose Reinforcing Plate	1
28	201282	Thread Ring	1
29	201284	Knob Standard	2
30	201287	Strike Box	1
31	201288	Strike Plate	1
32	201289	Shoe Retainer Cap	1
33	201290	O Ring 1 in. I.D.	1
34	201291	Screw, Philips Combination No. 8	4
35	201293	Clip	3
36	201298	Key DF 59 and Key Ring	2
37	201301	See Assembly Chart	1
38	201303	See Assembly Chart	1
39	201304	Clutch Backing Plate	1
40	201335	Combination Change Stud	1
41	201336	Combination Change Sleeve	1
42	201343	Front Knob Retainer Clip	1
43	201344	Inside Knob Retainer Clip	1
44	201345	See Assembly Chart	1
45	201350	Chamber Release	1
46	201351	Chamber Release Cam	1
47	201352	Bushing, Passage Set	1
48	201358	Connecting Bar	1
49	201362	Washer	1
50	201365	Knob Insert Best/Falcon	1
51	201366	Shoe Spring	2
52	201367	Turn Knob	1
53	201397	See Assembly Chart	1
54	201398	See Assembly Chart	1
55	201414	Knob Insert Russwin/Corbin	1
56	201416	See Assembly Chart	1
57	201417	See Assembly Chart	1
58	201422	Key DF 5 for Passage Control	2
59	201425	Shaft Self-Aligning Bushing 1000-2,4 & 6	1
60	201426	See Assembly Chart	1
61	201427	See Assembly Chart	1
62	201429	Combination Chamber	1
63	201430	See Assembly Chart	1
64	201431	See Assembly Chart	1
65	201432	See Assembly Chart	1
66	201433	See Assembly Chart	1
67	201434	See Assembly Chart	1
68	201435	See Assembly Chart	1
69	201436	Cross Pin .075 Dia.	1
70	201437	Cross Pin .100 Dia.	1
71	201438	Back Plate	1
72	201439	See Assembly Chart	1
73	201440	Retainer Ring 3/8 I.D.	1
74	201441	Retainer Ring 1 1/4 I.D.	1
75	201442	Screw Stud 1/4-20 X 1 5/8	1
76	201443	Screw 8-32 X 2 1/2	2
77	201444	Screw Philips 6-32 X 3/8 F.H.M.S.	6
78	201445	Split Washer No. 8	17
79	201446	Screw Philips 6-32 X 1/4 R.H.M.S.	4
80	201447	Screw Philips 6-32 X 5/16 R.H.M.S.	2
81	201448	Screw Philips 6-32 X 3/8 R.H.M.S.	3
82	201449	Screw Philips 8-32 X 1/4 R.H.M.S.	4
83	201450	Screw Philips 8-32 X 1/2 R.H.M.S.	4
84	201451	Screw Philips 8-32 X 3/4 R.H.M.S.	4
85	201490	See Assembly Chart	1
86	201496	See Assembly Chart	1
87	201497	See Assembly Chart	1

ASSEMBLY CHART

ASS'Y NO.	INCLUDING ITEM NUMBERS	ASSEMBLY DESCRIPTION
200080	4, 36	Control Lock Assembly Combination Change (DF-59)
201048	19, 21, 22, 23, 24, 28, 32, 33, 43, 51, 74, 78, 79	Cylindrical Drive Unit Assembly
201050	14, 69, 70	Shaft Assembly Standard
201093	15, 50	Knob Assembly Key Override for Best/Falcon
201155	3, 4, 9, 36, 75	Inside Trim Plate Assembly
201267	20, 69, 70	Shaft Assembly, Key Override
201301	16, 19, 21, 22, 23, 24, 32, 33, 40, 41, 43, 49, 51, 71, 73, 74, 78, 83	Back Plate Assembly
201303	26, 27, 28	Rose Assembly
201345	22, 32, 33, 43	Inside Drive Sleeve Assembly
201397	9, 34, 48, 58, 70, 86	Key Actuator Assembly for Passage Set
201398	9, 34, 48, 52, 70	Turn Knob Actuator Assembly for Passage Set
201416	55, 56	Knob Assembly Key Override for Corbin/Russwin
201417	20, 42, 55, 56, 59, 65	Conversion Assembly 1000-1 to 1000-2C
201426	1, 2, 8, 13, 14, 29, 42, 62, 63, 66, 67, 78, 80, 81	Front Plate Assembly 1000-1
201427	1, 2, 8, 13, 15, 20, 42, 50, 59, 62, 63, 64, 67, 78, 80, 81	Front Plate Assembly 1000-2B
201430	6, 7, 63	Clutch Sub-Assembly
201431	11, 64	Sleeve Drive Assembly Key Override for Best/Falcon
201432	11, 65	Sleeve Drive Assembly Key Override for Corbin/Russwin
201433	11, 66	Sleeve Drive Assembly Standard
201434	39, 67	Clutch Cover Assembly
201435	5, 10, 35	Link Assembly
201439	40, 41, 49, 71, 73	Back Plate Sub-Assembly
201490	15, 20, 42, 56, 59, 64	Conversion Assembly 1000-1 to 1000-2B
201496	58, 86	Control Assembly for Passage Set (DF-5)
201497	1, 2, 8, 13, 15, 20, 42, 55, 56, 59, 62, 63, 65, 67, 78, 80, 81	Front Plate Assembly 1000-2C

Figure 18.40 Parts identification and assembly chart for the Unican 1000 lock unit. (*Simplex Access Controls Corp.*)

Figure 18.41 Preso-Matic push-button combination deadbolt lock. (*Preso-Matic Lock Company*)

Figure 18.42 Deadlatch version of the Preso-Matic lock. (*Preso-Matic Lock Company*)

Figure 18.43 Instant exit feature and also locking ability with just a touch of a button. (*Preso-Matic Lock Company*)

The unit can also be used for businesses that rely on electric strikes operated from remote locations. The Trine model 012 electric strike unit is given a unique configuration by the Preso-Matic factory to accept the deadlatch bolt.

The Preso-Matic lock has several combination possibilities, including four-digit combination, seven-digit combination, and master combination (for use with the four-digit combination version).

The *four-digit combination* provides a maximum of 10,000 possible combinations; the *seven-digit combination* can give up to 10 million possible combinations. The last option adds a master combination to the individual four-digit combination. All locks of a given system will open for one six-digit master combination, but each unit within the system can still be opened by its own four-digit combination. The master combination variations allow for security protection and flexibility that are not available with other pushbutton keyless combination locks.

The numbers of the combination for each lock unit are determined by the two *combination slides* (Figure 18.44) that are inserted into the unit. To change the combination, you simply remove the inside cover plate, turn two spring clips clear of the slot in the lock housing, lift out the combination slides, and slip in new ones.

Installation. All the Preso-Matic locks are carefully designed and engineered to aid you in quick, easy, trouble-free installation. There are only three basic steps:

1. Bore the opening for the lock and latch (Figure 18.45) using the required template that comes with the unit (Figures 18.46 and 18.47).

2. Mortise the door edge for the bolt face and insert the bolt housing and tighten the screws (Figure 18.48).

3. Insert the lock and tighten only one screw, connecting the bolt housing to the lock and making the unit operable (Figure 18.49). Attach the cover plate on the inside of the door. Installation is complete.

For volume installations in large construction projects or numerous prefabricated doors, you should use the Preso-Matic round face, drive-in bolt, which installs in seconds and saves time.

Three basic installation layouts are available, as shown in Figure 18.50. Figure 18.51 is a quick summary of the various lock models. Figure 18.52 provides an exploded view of the lock and part identification.

Figure 18.44 Combination slides used with the lock unit. (*Preso-Matic Lock Company*)

Figure 18.45 Preso-Matic installation template, 2⅜-inch backset.

Figure 18.46 Preso-Matic installation template, 3-inch backset.

TEMPLATE

TOP

5 3/8"

1" HOLES

1 7/8"   3" BACKSET

FOLD ALONG THIS LINE.   FOR DOORS HINGED ON LEFT

BORE 1" HOLE
PARALLEL TO
SIDES OF
DOOR.

LOCATE
CENTER ON
THIS LINE.

PLACE ON OUTSIDE OF DOOR

Figure 18.47 Preso-Matic installation, Step 1.

Figure 18.48 Preso-Matic installation, Step 2.

Figure 18.49 Preso-Matic installation, Step 3.

MODEL	LOCK TYPE	BOLT PROJECTION	BACKSET	NOTES
8101	Deadbolt	$5/8$"	$2^3/8$" ($2^3/4$ also available)	Manual locking. Built-in night latch button. Hardened steel free-turning bolts. Manually locks.
LT8102	Deadbolt	1"	3"	Same
8103	Deadlatch	$1/2$"	3"	Night latch button. Locks automatically. Inswinging doors only.
8103A	Deadlatch	$1/2$"	3"	Same, except without night latch button
8200	Deadlatch	$1/2$"	3"	Hardened steel latch bolt. Locks automatically. Night latch button
8200A	Deadlatch	$1/2$"	3"	Same, except without night latch button.

OPTIONS:
1. Change of combination slides for a 7 digit code available.
2. Lock shields.
3. Stay open function; keeps the deadlatch in an unlocked position.
4. Door stop strike plate; deadlatch models only. Case hardened; ideal for double door installation.
5. Extra wide strike plate for metal doors ($2^1/2$" × $2^1/2$"); case hardened.
6. Heavy duty ANSI electric door opener. 24 VAC. for normally locked units.

Figure 18.50 Installation layout variations. (*Preso-Matic Lock Company*)

	A	B	C	D	E	F
Model 8101 2⅜″ BACKSET	1⅞	5⅜	1	1½	¹³⁄₁₆	3⅜
Model 8101 2¾″ BACKSET	1⅞	5⅜	1	1⅞	¹³⁄₁₆	3¾
Model LT8102 3″ BACKSET	1⅞	5⅜	1	2 ¹⁄₁₆	¹³⁄₁₆	3¹⁵⁄₁₆
Models 8103, 8103A, 8200, 8200A	1⅞	5⅜	1	2 ¹⁄₁₆	¹³⁄₁₆	3¹⁵⁄₁₆

Figure 18.51 Overview of the various lock models. (*Preso-Matic Lock Company*)

Exploded View Parts List:
1. Large Headed Screw
2. Spring Clips
3. Unlocking Lever
4. Unlocking Lever Pin
5. Bolt Actuator
6. Bolt Housing Casting
7. Multiplier Disc
8. Bolt
9. Bolt Spring
10. Pull Pin
11. Rocker Arm
12. Rocker Arm Pivot Clip
13. Flat Spring
14. Unlock Button
15. Lock Button
16. Inside Cover Plate
17. Long thin spring
18. Pull Plate
19. Combination slide cover plate
20. Combination slides
21. Lock body cover plate
22. Neutralizing spring
23. Number button springs
24. Flat inclined plates on reset button
25. Reset button
26. Number buttons
27. Lock body
28. Deadbolt & Housing

Figure 18.52 Exploded view of the Preso-Matic pushbutton combination lock and parts identification. (*Preso-Matic Lock Company*)

Electrical Access and Exit Control Systems

Electrically operated release latch strikes and locks are easily installed in place of standard units to provide remote controlled access for a door. These units are not new to business and industry and really shouldn't be that new to the locksmith—but they are. Why? Because, excepting a few instances in large cities or industrial regions, the average locksmith doesn't get involved with such units. Since it is "electrical," the job is passed to an electrician. By doing so, the locksmith loses valuable business.

This chapter discusses some electrically operated locks and release latch strikes, the circuitry behind the various types of units, where they can be installed, and the potential of such units.

Electric Release Latch Strikes

Electrically actuated release latch strikes are more common than locks. These units normally can be reversed for either right- or left-hand doors. They operate on low voltage and the strikes fit a hollow jamb channel as shallow as 1⅝ inches (1¾ inches for mortise latches) or they can be, if necessary, installed in wooden jambs.

Electrical release strikes are gaining popularity among small and large businesses. They come in a variety of sizes (Figure 19.1) to meet varying needs and requirements. There are many electrical strikes available on the market. While all of them meet minimum standards, prior to your purchase of them you should still consider the manufacturer and the purpose for which the units were developed.

Since this is a rapidly-growing area of access control (security), it is important that you receive that latest and best information available. To obtain the following information, I contacted Adams Rite, the manufacturer considered the nation's leader in the electrical release latch strike field, Adams Rite. I am extremely grateful for Adams Rite's assistance.

Figure 19.1 Various types and sizes of electric release latch strikes. (*Adams Rite Mfg. Co.*)

First and foremost in the field is Adams Rite, whose electrical strikes are architecturally sound and designed to add increased security features to areas requiring a certain amount of traffic control. When properly powered and installed, these electrical release strikes will provide a very long, maintenance-free life.

The fact that electrical power is used for the operation of the strike in no way makes the unit an electrical appliance, as some people tend to think. With a few exceptions, the electrical circuitry unique to these units are designed to meet the needs of the strike, not the other way around. The hardware specifier (that's you) should select the strike and expect the electrician (which could be you) to supply the power at the point of installation. In a few cases an existing circuit can be used, so the strike is selected and installed to it. In selecting strikes, two areas of consideration must be followed: electrical and hardware.

Electrical considerations

First, determine the duty of the unit. Is operation to be intermittent or continuous? If the door is normally locked and released only momentarily from time to time, it is *intermittent*. In some cases, the strike is activated (unlocked) for long periods, so duty is *continuous*. A more rare requirement is *reverse action*, in which the strike is locked only when its current is switched on.

Intermittent duty. For a normal intermittent application, the electrical strike will have to be 24 Vac. This alternating current call-out gives enough power for almost any entrance, even one with a wind-load situation. Yet the low voltage range is below that requiring UL or building code supervision. At this voltage, good, reliable transformers are available.

Continuous duty. When a continuous duty application is required, additional components are added to the unit at the factory.

Reverse action. A long period of unlocking can also be obtained by using a reverse action strike. This might be required to provide the same service as a continuous duty strike, but preserves current because it is on battery dc power. It could also be used to provide a fail-safe unlocked door in case of a power failure or building emergency (such as a fire) where quick, guaranteed unlocked doors are required. *Note:* Reverse action strikes, when ordered by the locksmith for installation, must specify the hand of the door on the order. These types are not reversible, as are some other strikes.

Monitoring. If a visual or other signal is required to tell the operator that the electric strike is doing its job properly, then a monitoring strike will be required. Two sensor/switches are added: one is activated by the latchbolt's penetration of the strike, and the other by the solenoid plunger that blocks the strike's release.

Transformer. Low voltage is necessary for proper electric strike operation, and this is obtained by the use of transformers which are stepped down from the normal ac power voltage to 12, 16, 24, or other lower voltage. Whether or not you perform the electrical side of the installation (when wires are required to be run) three electrical specifiers are necessary:

- Input voltage (110—standard household current)
- Output voltage (12, 16, 24, etc.—from the transformer unit)
- The capacity of the transformer, called *volt amps* (output voltage times the output amperes)

Skimping on the capacity of the transformer to save a few dollars will underpower the door release and is likely to bring in complaints of "poor hardware." Intermittent duty electric strikes draw from 1½ to 3 amps; a continuous duty model draws less than 1 amp.

In any application where the current draw of the strike must be restricted to less than 1 amp, regardless of the duty, a current limiter which stores electrical energy to relieve high use periods is available.

Wiring. The wiring must carry the electrical power from the transformer through the actuating switch to the door release. It must be large enough to minimize frictional line losses and deliver most of the input from the transformer to the door release. A small diameter garden hose won't provide a full flow of water from the nozzle, particularly if it's a long run; neither will an undersized wire carry the full current.

Electrical troubleshooting. When insufficient electrical power is suspected in a weak door release, make a simple check. Measure the voltage at the door release while the unit is activated. If the voltage is below that specified on the hardware schedule, the problem is in the circuit—probably an under-capacity transformer, if the current length is short. A long run may mean both a transformer and a wire problem.

Hardware

Hardware is the second factor in selecting a strike. There are many considerations.

Face shape (sectional). Electrical strikes from Adams Rite have a flat face. Two exceptions have a radius-type to match the nose shape of a paired narrow stile glass door.

Face size. The basic Adams Rite size conforms to ANSI standards for strike preparation: 1¼ inches × 4⅞ inches. However, two other sizes are offered to fill or cover existing jamb (or opposing stile) preparation from a previous installation of another unit, such as an MS deadlock strike or a discontinued series electric strike.

Face corners. Strikes are available with two types of corners: round for installation in narrow stile aluminum where preparation is usually done by router, and square for punched hollow metal ANSI preparation or for wood mortise installation.

Jamb material. Strikes with vertically mounted solenoids are designed to slip into hollow metal stile sections as shallow as 1⅝ inches. Horizontal solenoids for wood jambs require an easily bored mortise of 3⅛ inches.

Previous strike preparation. If the jamb was previously fitted with a now-discontinued strike, such as a Series 002 electrical strike, the new strike unit to be installed—perhaps an Adams Rite 7510—will fit in with only minor alterations. In this example, you would increase the lip cutout by ¼ inch at each end. In the case where a hollow jamb was originally prepared to receive the bolt from another type deadlock, another latch such as the 4510 series would have to be substituted. In such a case, the radius or flat jamb would cover the old strike cutout.

Lip length. The standard lip on all basic Adams Rite electric strikes accommodates a 1¾-inch thick door that closes flush with the jamb. Where the door/jamb relationship is different, a long lip can be added.

Compatible latch. Certain electric strikes will mate with all latches in a given series. The key-in-knob latches and mortise latches are made to mate properly with the various electric strikes available.

Handing of the strike unit. All standard operation strikes are unhanded; thus, they can be installed for either a right- or a left-hand door. However, reverse action (locked only when the current is on) units require that you specify the hand when you place the order for the unit.

To repeat: understand that electrical strikes are not just electrical appliances; they are hardware. In the case of Adams Rite strikes, they are made by hardware people, because they know that the hardware has to be carefully matched to its application.

Also, realize that when you order a unit for installation, you need to check with the local electrical codes to ensure that you can actually install it. Some jurisdictions may require an electrician to do the job.

Installation and procedures

The following section discusses and illustrates some of the procedures required to install a variety of strike units.

Aluminum jam installation (Series 750)

1. Prepare the door jamb per Figure 19.2.
2. Install the mounting clips to the jamb using 8-32 × ⅜ screws and pressed metal nuts. Leave the screws slightly loose to permit easy alignment of the base assembly and clip.

Figure 19.2 Jamb specifications and installation details for a radial (curved) aluminum door jamb. (*Adams Rite Mfg. Co.*)

3. Spacers are provided to assure flush final assembly of the face plate and jamb. Add one or more spacers between the jamb and mounting clip when the face plate extends beyond the jamb. When the face plate sets inside the jamb, you must add spacers between the mounting clip and the electric strike case. See detail "A" in Figure 19.2. To attach spacers to the mounting clip, remove the protective coating from the spacer and press to the desired mounting clip surface. Make sure the clearance hole in the spacer aligns properly with the hole in the mounting clip.

4. Using the wire nuts, connect the wires coming from the unit to the wires coming from the low voltage side of the transformer.

5. Insert the electrical release into the jamb by tipping the solenoid coil into place behind the mounting clip, then drop the unit onto the clips.

6. Attach the unit with the #10 screws and lock washers provided.

7. Attach the face plate using the screws furnished (8-32 × ¼ inch).

8. Secure the screws holding the mounting clips to the jamb.

Figure 19.2 also shows the dimensions of the electric door release that would be installed in the aluminum jamb.

Replacing an obsolete unit with an electrical strike (model 7510 wood jamb)

1. Prepare the jamb to the dimensions shown in Figure 19.3. Take care to mortise the cut to clear the solenoid.

2. Using the wire nuts provided, connect the wires from the 7510 unit to the wires from the low-voltage side of the transformer.

3. Insert the electric release into the jamb. Attach this with #10 flat head wood screws.

Replacing the obsolete unit. Unit 002 is used as an example.

1. Remove the 002 unit and disconnect the low-voltage wires. Do not remove the existing mounting clips.

2. Enlarge the jamb cutout by removing ¼ inch from each side.

3. Using the wire nuts, connect the 7510 wires to the low-voltage side of the transformer.

4. For a flush installation, provide a ¹⁄₁₆-inch spacer between the mounting surface and the face plate.

5. Insert the electric door release into the jamb, and attach with screws.

Figure 19.3 also shows the dimensions of the 7510 electrical door release unit.

Figure 19.3 Dimensions and installation details for unit to be mounted in a wood frame. (*Adams Rite Mfg. Co.*)

Installing an electric strike to meet ANSI specifications.   Model 7520 is used in the example. Note the molded strip down the center of the door frame in the illustration. Installation is as follows:

1. Prepare the metal jamb to the dimensions shown in Figure 19.4.
2. Spacers are provided with the unit to assure the flush mounting of the face plate to the jamb. To use the spacer(s), remove the protective cover and press them to the mounting bracket. Make sure the clearance hole in the spacer aligns with the tapped hole in the bracket.
3. Connect the wires from the unit to the transformer wires.
4. Insert the electric release into the jamb by tipping the solenoid coil into place behind the mounting bracket. Then drop the unit onto the mounting brackets. Attach the unit with the screws furnished.
5. Attach the face plate.

Installing an electric strike for key-in-knob latch to ANSI specifications.   The example used is an Adams Rite Model 7540 Unit.

1. Prepare the door jamb as shown in Figure 19.5.
2. Install the mounting clips using the screws and pressed metal nuts. Leave the screws slightly loose to permit easy alignment of the case assembly and clip.
3. Use the provided spacers as necessary.
4. Connect the wires from the 7540 unit to the low-voltage wires from the transformer.
5. Insert the electric release into the jamb and attach with the #10 screws provided.
6. Attach the face plate and secure the screws holding the mounting clip to the jamb.

Installing an electric strike in a wooden jamb

1. Prepare the wood door jamb to the dimensions indicated in Figure 19.6.
2. Connect the wires from the unit to the transformer wires.
3. Insert the strike into the jamb and attach with the #10 wood screws.
4. Attach the face plate.
5. Figure 19.7 provides the technical dimensions of the strike unit to be installed.

Circuitry

Figure 19.8 illustrates the typical wiring diagram for an audible (ac) operating strike unit. Figure 19.9 is the same unit with a rectifier used to provide silent (dc) operation of the strike unit.

Figure 19.4 Jamb preparation and dimension data for a metal frame that has a molded strip down the frame center. (*Adams Rite Mfg. Co.*)

Figure 19.5 Installation details for a key-in-knob latch unit. (*Adams Rite Mfg. Co.*)

Figure 19.6 A wooden door jamb preparation. (*Adams Rite Mfg. Co.*)

1.065
(27.05)

3.094
(78.59)

1.59
(40.39)

.656
(16.66)

1.687
(42.85)

1.250
(31.75)

.267
(6.78)

.750(19.05)

3.370
85.60

4.870
(123.70)

1.50
(38.10)
2.60 reverse
action only (66.04)

1.641
(41.68)

DIMENSIONS In inches and
(millimeters)

Figure 19.7 Strike unit dimensions for wood door jamb installations. (*Adams Rite Mfg. Co.*)

You might have an instance where the installation requires a monitoring signal operation. Figure 19.10 provides a typical wiring diagram. Many monitoring arrangements are available. The green light indicates that the latchbolt is in the strike and that the solenoid plunger is blocking the strike. The yellow light indicates the latchbolt is in the strike but is unblocked, allowing access. The red light signals that the latchbolt is out of the strike.

Troubleshooting Adams-Rite electric strike latch units

You have seen several of the more common types of electrical strike release units. What do you do if you have trouble with a unit after you install it?

Figure 19.8 Typical wiring diagram for an audible (ac) operation. (*Adams Rite Mfg. Co.*)

Figure 19.9 Same circuitry as shown in Figure 19.8 but with a rectifier added for silent (dc) operation. (*Adams Rite Mfg. Co.*)

In this, one of many possible monitoring arrangements. The green light indicates that latch bolt is in strike and solenoid plunge is blocking strike. Yellow light indicates latch both is in strike but strike is unblocked, allowing access. Red light signals that latch bolt is out of strike.

Figure 19.10 Wiring for a monitor signal operation. (*Adams Rite Mfg. Co.*)

Accurately checking an electrical circuit, just as any other job, requires specific tools. You need a *good* 20,000 volts per ohm volt-ohm-milliammeter. The Simpson model #261, one of several currently available, costs about $60, but it is money well spent if you will be installing and maintaining electric strike latch units.

First, read the volt-ohm-milliammeter (VOM) instructions and practice on simple low-voltage circuits to ensure that you properly understand the operating instructions and can correctly read the results on the meter.

Checking voltage

1. Zero the pointer.
2. Turn off the power to the circuit being measured.
3. Set the function switch to the correct voltage to be measured (+dcC or ac).
4. Plug the black test lead into the common (–) jack; plug the red test lead into the (+) jack.
5. Set the range selector to the proper voltage scale. *Caution:* It is very important that the selector be positioned to the nearest scale voltage to be measured.
6. Connect the black test lead to the negative side of the circuit and the red lead to the positive side. This is applicable to dc circuit only. Turn the power on to the circuit being tested. If the pointer on the VOM moves to the left, the polarity is wrong. Turn the switch function to –dc and turn the power back on. The pointer should now swing to the right for the proper reading on the dc scale. For an ac circuit, the connections are the same except as noted in step 3. You don't have to worry about polarity in the ac circuit.
7. Turn the circuit off before disconnecting the VOM.

Checking current (dc)

1. Zero the pointer.
2. Turn off the power to the circuit being tested.
3. Connect the black test lead to the –10A jack and the red test lead to the +10A jack.
4. Set the range selector to 10 amps.
5. Open the circuit to be measured by disconnecting the wire that goes to one side of the solenoid. Connect the meter in series—that is, hook the black lead to one of the disconnected wires and the red lead to the other wire.
6. Turn the power on to the circuit and observe the meter. If the pointer moves to the left, reverse the leads in the jacks.
7. Turn the power back on to the circuit and read amperage on the dc scale.
8. Turn the VOM off before disconnecting.

Checking for line drop.   Measure the line drop by comparing voltage readings at the source (the transformer's secondary or input side) with the reading at the electric strike connection.

Finding circuit shorts in a hurry. The VOM is the most reliable instrument for detecting a circuit short. This is accomplished by setting up the VOM to measure resistance.

1. Set the range switch at position R × 1.
2. Set the function switch at +dc.
3. Connect the black test lead to the common (–) jack and the red lead to the (+) jack.
4. Zero the pointer by shorting the test leads together.
5. Connect the other ends of the test leads across the resistance to be measured—in the case of a solenoid, one of the test leads to one coil terminal and the other test lead to the other terminal.
6. Watch the meter. If there is no movement of the pointer, the resistance being measured is open. If the pointer moves to the peg on the right side of the scale, the resistance being measured is shorted closed. If you get a reading in between these two extremes, it is very likely the solenoid is okay.

Problem solving

The following are solutions to several common problems.

The strike will not activate after installation

- Check the fuse or circuit breaker supplying power to the system.
- Check that all wiring connections are securely made. When wire nuts are used, ensure that both wires are firmly twisted together for good electrical transfer between the wires.
- Check the solenoid coil rated voltage (as shown on the coil label) to make sure that it corresponds to the output side of the transformer within ±10%.
- Using the VOM, check the voltage at the secondary (output) side of the transformer.
- Using the VOM, check the voltage at the solenoid. This assures you that there are no broken wires, bad rectifiers, or bad connections.
- Check the coil for a possible short.

The transformer overheats

- Make sure that the rated voltage of the transformer and the rated voltage of the coil correspond within ±10 percent.
- Make sure the VA rating is adequate. For all Adams-Rite units, it is recommended you see 40 VA. 20 VA is the absolute minimum, and you could experience transformer heating in even moderate use applications.

The rectifier overheats

- The rectifier is wired wrong. This means the overheating is of a temporary nature (a few milliseconds) and then it's burned out.

- There are too many solenoids being supplied by a single rectifier and you are pulling more current through it than the system diodes are rated for.

The solenoid overheats

- The coils used in an electric strike latch, when used as rated (continuously or intermittently by pulsing once per second), gave a coil temperature rise rating of 149°F (65°C) *above* ambient. To get the exact temperature, add 149°F plus 72°F (ambient), which equals 221°F. The coil insulation is rated at 266°F (130°C). Regardless of whether the coil goes to 221°F or 266°F, it is too hot to keep your fingers on!

- With the above in mind, it is fair to say the vast majority of intermittent duty units never see that kind of use. If a coil gets extremely hot on very short pulses at two- or three-second intervals, you either have the wrong coil or the wrong transformer output. The same is basically true for continuous duty coils. If the coil temperature exceeds the ratings, it has to be because the coil voltage or the transformer are improperly coordinated.

- If you set up the meter as if testing for a short (see previous instructions) and obtain the exact resistance, you can compare to the specifications found in Figure 19.11; this becomes another way of knowing you have the correct coil. Figure 19.11 also provides all the power data for the various electric strikes from Adams Rite that you might ever be required to install on various premises.

Determining the proper wire size

Some people—especially those not trained in the various phases of proper electricity—assume that a low voltage means you can use the smallest and cheapest wire size available. This is certainly not the case. The size of the wire is dependent on the use, the amount (voltage) of electricity going through the wire, the distance the wire runs, and the number of strikes to be used on the wire run.

To determine the correct wire size, obtain the wire size factor from Table 1 of Figure 19.12 and multiply this number by the distance to the strike (in feet). Take the resulting number in *circular mills* (CM) to determine the wire size required from Table 2 of the Figure.

Example 1: A 24 volt ac intermittent (24 Vac-int) is to be mounted 60 feet from the transformer. The wire size factor for the unit is 34. Multiply 34 times 60; this equals 2040. On Table 2, 2040 corresponds to a wire size of 16 gauge.

Example 2: What is the maximum distance that you could mount the same 24 Vac-int using a 20 gauge wire? From Table 2, the largest CM number corresponding to 20 gauge is 2000. Divide 2000 by the wire size factor for 24 Vac-int (34) to get 29. The maximum distance is 29 feet.

STRIKE VOLTAGE, CURRENT AND DUTY:	6 VDC - INT	12 VDC - INT	24 VDC - INT	48 VDC - INT	12 VDC - CONT	24 VDC - CONT	16 VAC - INT	24 VAC - INT	48 VAC - INT
COIL RESISTANCE (Ohms) Ω	1.0	3.5	15.0	61.0	15.0	61.0	7.4	3.5	15.0
PEAK INSTANTANEOUS CURRENT (Amps)*	5.8	2.7	1.42	.70	.75	.36	1.6	7.0	3.2
CONTINUOUS OR HOLD CURRENT (Amps)*	5.8	2.7	1.42	.70	.75	.36	.70	3.72	1.9
PEAK INSTANTANEOUS POWER (Watts)	34.8	32.4	34.2	33.6	9.0	8.6	25.6	168.0	154.0
CONTINUOUS OR HOLD POWER (Watts)	34.8	32.4	34.2	33.6	9.0	8.6	11.3	89.2	95.0
TURN-ON SETTLE TIME (Milliseconds)*	40	11	12	14	30	30	8	8	8

* Readings taken from oscilloscope. Photographs made.

Figure 19.11 Power data for electric strike units. (*Adams Rite Mfg. Co.*)

STRIKE USED	WIRE SIZE FACTOR
6 VDC — INT	128
12 VDC — INT	50
24 VDC — INT	13
48 VDC — INT	3.5
12 VDC — CONT	14
24 VDC — CONT	3.5
16 VAC — INT	10
24 VAC — INT	34
48 VAC — INT	9

TABLE 1

CM	WIRE GAGE
1000 AND BELOW	20
1001 — 1600	18
1601 — 2550	16
2551 — 4100	14
4101 — 6500	12

TABLE 2

Figure 19.12 Tables for determining what size is required based on type of electric strike unit used. (*Adams Rite Mfg. Co.*)

If two or more strikes are to be actuated simultaneously on the same circuit, multiply the wire size factor by the number of strikes. Then, multiply by the distance to the farthest strike to obtain the CM number, and the wire gauge required for the installation.

Electric Door Openers

The Trine Consumer Products Division of the Square D Company is another major supplier of electric door openers. Trine units are available for standard wood and metal doors and narrow stile doors, such as those obtainable from Adams-Rite (Figure 19.13). Fourteen different models are currently available for such diverse application as entrance doors to apartments, banks, institutions, industrial buildings and complexes, offices, and interlocking door control.

Installed in the door jamb in place of the lock strike or in conjunction with various other locksets, these units offer the ability to lock or unlock a door from a remote control station. This variety of electric strike can be used with most standard brands of locksets in both mortise and surface mount types.

No. 001
Mortise
Type
Electric
Door
Opener

No. 002
Mortise Type
"Adams-Rite"
Electric
Door
Opener

No. S005
Mortise Type
Electric Door Opener

No. 012
ANSI UL Listed
Mortise Type
Electric
Door Opener

Figure 19.13 Electric door openers come in a variety of sizes and shapes to fit different installation application needs. (*Trine Consumer Products Division, Square "D" Company*)

ELECTRICAL OPERATING CHARACTERISTICS
Average Current Ratings

ELECTRO-MAGNET TYPE (normally LOCKED . . . UNLOCKED when energized)					
AC Voltage	*Coil-Ohms Resistance	Current Draw	DC Voltage	*Coil-Ohms Resistance	Current Draw
10	2.0	1.95	6	2.0	3.0
16	2.0	2.0			
**24	15.0	.66	**24	15.0	2.0
48	73.0	.18	48	73.0	.66

* Plus or minus 10%.
** Suitable for CONSTANT DUTY OPERATION.

SOLENOID TYPE Available (normally LOCKED) and (normally UNLOCKED)					
AC Voltage	*Coil-Ohms Resistance	Current Draw	DC Voltage	*Coil-Ohms Resistance	Current Draw
FOR INTERMITTENT USE					
24	47.0	.42	24	119.0	.26
			48	200.0	.10
FOR CONSTANT DUTY OPERATION					
24	119.0	.26	24	119.0	.26

*Plus or minus 10%

Figure 19.14 Electrical operation characteristics for Trine units. (*Trine Consumer Products Division, Square "D" Company*)

Remember that electric strike units have two systems of operation. For unlocking doors (normally in the locked position), it opens when energized. For locking doors (normally unlocked) the door is locked when the unit is energized.

For unlocking doors, the units come in standard voltages: 3 to 6 Vdc and 8 to 16 Vac. These units are electromagnetic coil types. For locking doors they come in standard voltage of 24 Vac, with a 47 ohm solenoid, and 24 Vac 119 ohm solenoid. These units are suitable for constant duty operation without overheating or burnout.

Both types of units are fail-safe; they will automatically go to the unlock position in the event of power failure, which assures an unlocked passageway in case of emergency.

Figure 19.14 provides the electrical operating characteristics for both electromagnetic and solenoid units. Figures 19.15 through 19.17 illustrate several models commonly installed by locksmiths.

Figure 19.15 Trine model 014 for key-in-knob locksets. (*Trine Consumer Products Division, Square "D" Company*)

Figure 19.16 Solenoid coil unit for high-intensity use. (*Trine Consumer Products Division, Square "D" Company*)

Figure 19.17 Model 003 with a magnet coil for normal use; this unit is used with rim-type locksets. (*Trine Consumer Products Division, Square "D" Company*)

Figures 19.18 and 19.19 are specification guide details relating to the actual full-size opening of the various types of Trine electric door opener latch cavities. Remember when you install one to check the lock latch width and throw to the cavity size of the door opener selected, allowing for a minimum ¹⁄₃₂-inch clearance at all points. For lock latch height for the unit installed, always check the detailed drawing provided with the lock latch to ensure that you have properly cut the cavity and not over- or under-cut. For the lock latch heights, see the detail drawings in Figures 19.20 and 19.21. As an example, a typical installation detail using the Trine Model 001 is shown in Figure 19.22. Jamb preparation for the installation of an electric strike in a metal door frame is in Figure 19.23.

During jamb preparation, a neat cutting job is required for good appearance, so take special care to ensure the height dimension is adequate for a loose fit insertion in the cutout. A tight fit here will cause the keeper to bind in the case and result in poor performance.

Figure 19.18 Size specifications for various latches. (*Trine Consumer Products Division, Square "D" Company*)

The dimension indicated by an asterisk must be determined by the installer or the jamb fabricator; it is dependent on the lockset used. This dimension establishes the relationship between the keeper and the bolt. Incorrect location of the strike will prevent the door from shutting tightly or cause excessive binding between the keeper and the bolt. Leave approximately ⅟₃₂ inch free space between the keeper and the bolt.

This is a simple installation, but the two strike wires must be attached securely by wires nuts to the supplied voltage lines. Be careful not to damage the wires when inserting the strike in the jamb cutout. This can be accomplished by hooking the wire outlet side under the jamb edge first, then pushing the other side of the strike in proper position.

Low-voltage electric door opener circuitry

Figure 19.24 illustrates and provides specifics for audible (ac) operation, while Figure 19.25 is for silent (dc) operation.

Figure 19.19 More latch cavity specifications. (*Trine Consumer Products Division, Square "D" Company*)

No. S005
Mortise type for Schlage
type (bored cylindrical) lock
sets. 6" wire leads.

No. 003
Narrow, surface type. For
mounting in conjunction
with rim type lock sets. 6"
wire leads.

Figure 19.20 Detail specification drawings for magnet coil units. (*Trine Consumer Products Division, Square "D" Company*)

While it seems impracticable, some businesses even use the automatic electric interlock type of circuitry for rooms that serve more than one consumer, such as a bathroom between two offices or separate businesses located in the same building. Figure 19.26 shows how the electric interlock works in such a situation. Such interlocks can also be used for air shower rooms, money counting rooms, x-ray and photographic rooms, radiation and biological labs, and other such installations.

In Figure 19.26, the door opener is normally unlocked, but it becomes locked when energized. The door indicator lights (DL-1/DL-2) can be used if a visual signal is required. The opening of either door operates the door switch (a mortise type, Trine #340 or surface type, Trine model #316). It also activates the opposite unlocked door opener, locking the door. When the door shuts, the opposite door returns to the unlocked position.

Pretend, for example, that the unit is used to lock a communicating bathroom. On entering the bathroom, the occupant turns the outer knob, unlocking the lockset. To lock the bathroom's doors, the occupant closes both doors and operates the wall switch to locked position, illuminating both DL-2 door indicator lights and locking both doors. Operating the wall switch to the unlocked position turns off both door indicator lights and unlocks both the doors. After leaving the bathroom, the occupant presses the pushbutton in the lockset, preventing anyone from the adjoining room from entering the occupant's room. An emergency release is obtained by installing a pushbutton at each door, which will unlock its own door.

No. 005
Standard Mortise type.
6" wire leads.

No. 006
Standard surface type. 6"
wire leads.

No. 007
Mortise type for A.N.S.I.
Standard strike cut-out (with
modification of jamb).
6" wire leads.

No. 004
Mortise type. For use with
locks with dead bolts.
6" wire leads. "Knock-outs"
top & bottom.

Figure 19.21 Specification details for high-intensity solenoid coil electric door opener units. (*Trine Consumer Products Division, Square "D" Company*)

Figure 19.22 Installation detail for model 001 magnet coil unit. (*Trine Consumer Products Division, Square "D" Company*)

Figure 19.23 Procedural details for a mounted door frame installation. (*Trine Consumer Products Division, Square "D" Company*)

This is only one type of situation where more than one door opener (always an RS type unit in such interlocking cases) is used. There are other instances where multiple door openers are required, each having its own pushbutton control switch and each one is operated off the same power line. Figure 19.27 illustrates the wiring of multiple door openers.

Figure 19.24 Low-voltage electrical circuit for an audible operation. (*Trine Consumer Products Division, Square "D" Company*)

Figure 19.25 Low-voltage circuitry for silent operation of unit. (*Trine Consumer Products Division, Square "D" Company*)

Figure 19.26 Automatic electric interlock for the operation of opposite side doors in an area servicing more than one user. (*Trine Consumer Products Division, Square "D" Company*)

Figure 19.27 Circuitry for multiple wiring of door openers.
(*Trine Consumer Products Division, Square "D" Company*)

Electrical ratings for solenoids for fail-safe (continuous duty) and standard (intermittent duty only) units are in Figure 19.28.

Due to the remote control applications for electrical strike units, there can be a considerable distance between the strike location and the actual control for that particular strike. Figure 19.29 shows the *minimum* wire sizes required for standard and fail-safe electrical strike unit voltage at various distances.

Troubleshooting Trine electrical strikes

No matter how carefully units are tested and re-tested at the factory, problems can still arise. These troubleshooting procedures concern the unit after installation, and they can relate to any electrical strike unit in use.

After installation, the strike hums but does not operate properly when the power is applied.   Probable cause: ac voltage has been supplied but the rectifier has not been installed between the transformer and the strike.

Correction: Install the rectifier per the wiring diagram in Figure 19.30.

ELECTRICAL RATINGS FOR SOLENOIDS	FAIL SAFE UNITS (CONT. DUTY)		STANDARD UNITS* (INTERMITENT DUTY ONLY)		
	24 VDC	12 VDC	24 VDC	12 VDC	48 VDC
RESISTANCE IN OHMS	115	33	16	4	90
WATTS SEATED	4.8	4.4	36	36	25
AMPS SEATED	.2	.37	1.5	3.0	.53

*Bridge rectifier may be supplied with any unit for AC operation.

Figure 19.28 Electrical ratings for intermittent use and constant operation for solenoid coil door openers. (*Trine Consumer Products Division, Square "D" Company*)

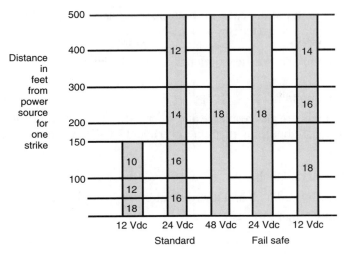

Figure 19.29 Electrical strike voltage (fail-secure mode of operation). (*Trine Consumer Products Division, Square "D" Company*)

After installation, the strike does not operate even if the power is left on for one minute (maximum time) and the strike is noticeably warm. Probable cause: (1) Static loading against the door is excessive due to weatherstripping, a warped door, wind loading on the door, or improper jamb preparation.

Corrections: (1a) Adjust the strike bolt relationship, so that some slight gap exists in the locked position. (1b) Balance air conditioning pressures on opposite sides of the door.

Probable cause: (2) Insufficient voltage is reaching the strike due to an inadequate supply wire size.

Corrections: (2) Provide sufficient sized service wire (Figure 19.29).

The keeper moves to fully open on power application, but does not return to the locked position when the door closes. Probable cause: If the deadbolt is extended but does not pull the keeper to the fully locked position on door closure, the return spring may need adjusting.

Correction: Deform the return spring slightly by forcing it toward the keeper.

Figure 19.30 Circuitry installation of a rectifier to alleviate a strike hum. (*Trine Consumer Products Division, Square "D" Company*)

The keeper binds in the strike case on both opening and closing. Probable cause: The cutout in the jamb is too narrow and it is forcing the strike case walls inward toward the keeper.

Correction: Increase the vertical dimension until the keeper operates freely.

Electric door openers for deadbolts

Electric door openers are also available and work extremely well for deadbolts. They can be used with just about any manufacturer's lockset. They include the following types:

- Mortise locks with deadbolt.
- Mortise locks with deadlatch and guarded latch.
- Individual deadbolt lock.
- Panic exit devices.
- Maximum security deadbolts (¾-inch throw).
- Release up to 1-inch throw deadbolts.
- With optional adapters, they may be used with cylindrical locks, alarm locks, or with Detex bolts.

These units will also allow the strike operation of mortise locks with both latchbolt and deadbolt release by the action of a single keeper. The keeper is held open after the door is opened if the deadbolt was in an extended position at the time of the door opening. The keeper automatically returns to the locked condition when the deadbolt reenters the strike. The door may also be closed with deadbolts extended. In strikes with options H, J, and K (Figure 19.31), the keeper is forced open by the extending latchbolt and springs closed to the locked position to accommodate the returning latchbolt. Other options are also shown in Figure 19.31.

Figure 19.32 illustrates two Trine unit model (015 and 016) specifications. Along with the basic unit for installation, a heavy duty transformer (Figure 19.33) is necessary to convert the primary voltage to the operating voltage.

Electrified Mortise Locks

The electrified mortise lock provides a wide number of additional installation sales and servicing possibilities, especially when access security controls for your customer are concerned. The electrified mortise lock (Figure 19.34) is another excellent, durable product that has minimal installation procedures yet offers long life and reliability to your customer.

This lock unit has the standard rugged lock case construction of stainless steel, plus a normal mortise strike. The inside knob is always free, while the outside knob is rigid except when the 24 Vdc internal solenoid is energized, which releases the outside knob and allows the latchbolt to retract. Like any other standard mortise lock, it is reversible, allowing for different hands.

Figure 19.31 Various strike options for electric door openers used with deadbolt locks. (*Trine Consumer Products Division, Square "D" Company*)

Figure 19.32 Technical specifications for two Trine door opener models that can be used with any manufacturer's lockset. (*Trine Consumer Products Division, Square "D" Company*)

Figure 19.33 Heavy-duty transformer for voltage conversion. (*Trine Consumer Products Division, Square "D" Company*)

Figure 19.34 Electrified key-in-knob lock. (*Alarm Lock Corporation*)

Installing the electrified mortise lock

Mark the door and jamb (Figure 19.35). Mark the horizontal centerline of the lock on both sides and the edge of the door. Mark the vertical centerline of the lock at the door edge and the vertical centerline on both sides of the door as measured from the door edge.

Mortise the door for the lock case and front (Figure 19.36). For the specific dimensions, refer to the installation template that comes with the lock unit.

Drill the holes as indicated on the template as shown in Figure 19.36. *Caution:* Check the lock for function, hand, and bevel before drilling. Mark and drill only those holes required for your function.

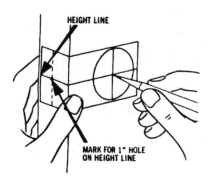

HEIGHT LINE

MARK FOR 1" HOLE
ON HEIGHT LINE

Figure 19.35 Marking the door for installation. (*Alarm Lock Corporation*)

BORE LATCH HOLE AND
MORTISE FOR LATCH FRONT

Figure 19.36 Bore and mortise for latch front. (*Alarm Lock Corporation*)

Remove the front plate and mount the lock in a mortised opening. Apply the trim (four trim variations are possible; see Figures 19.38 through 19.41 at the end of the installation instructions). Apply the cylinder (using a minimum of four turns); then, fasten with the cylinder clamp screw. Replace the front plate. *Caution:* Check the cylinder and lock for proper function before closing the door.

Install the strike (Figure 19.37). Position the strike on the door jamb as shown. Mortise the jamb and apply the strike. When using a box strike, mortise deep enough to allow the latchbolt and deadbolt to fully extend.

Reversing the lock and bevel

At times, you might be required to replace or remove and reinstall and lock on another door.

Figure 19.37 Strike installation details. (*Alarm Lock Corporation*)

1. Remove the cover from the lock.

2. Reverse the latchbolt and the auxiliary bolt.

3. Reverse the hubs.

4. To change the bevel, loosen the two screws on the top and bottom of the case. Adjust the proper bevel and tighten the screws.

5. Replace the cover. Be certain that all parts are in their proper place before fastening the exterior screws. *Caution:* When the hand of the lock is changed, make sure that the proper hand strike is also used. A left-hand lock requires a left-hand strike; a right-hand lock requires a right-hand strike.

Figures 19.38 through 19.41 show the various trims available for this unit. Figure 19.38 shows the heavy-duty screwless trim; Figure 19.39 shows the lever handle trim; Figure 19.40 is the medium duty trim with a Glenwood escutcheon; and Figure 19.41 is the lever handle trim with a Utica escutcheon.

Electrified Knob Locks

The electrified knob lock (Figure 19.42) by the Alarm Lock Corp. is a heavy-duty lockset with a key bypass in the outside knob. It has a deadbolting latchbolt and stainless steel trim for long-lasting exterior or interior use against the elements. The inside knob is always free for exit. The outside knob is rigid to prevent entry except when the internal solenoid is energized, which releases the outside knob to retract the latchbolt and open the door. The electrified knob lock has a continuous duty solenoid that uses 24 Vac power supply (360 milliamps).

Installation procedures

1. Mark the height line on the edge of the door 38 inches from the floor (this is the suggested height, which may vary according to individual needs). Position the centerline of the template that comes with the lock on the height line, and mark the center point of the door thickness and the center point for the 2⅛-inch hole through the door (Figure 19.43).

2. Bore the 2⅛-inch hole at the point marked from both sides of the door. Bore the 1-inch latch unit hole straight into the edge of the door from the center point of the height line. Mortise the latch front and install the latch unit (Figure 19. 44).

3. Remove the lock from the box and remove the inside trim by inserting the small end of a spanner wrench through the hole in the knob bearing sleeve. This hole faces the latch retractor. Press the knob catch to slide the knob off the spindle and remove the rose from the threaded sleeve (Figure 19.45).

4. Adjust for the door thickness by moving the outside rose either in or out until the indicator on the lock shows the proper thickness for the door. Make sure the rose clicks into place (Figure 19.46). The lock will fit on all doors 1⅜ inches to 2 inches thick. (If you need a lock for a thicker door, special order it from the company.)

5. Next, install the main unit (Figure 19.47). The main housing must engage with the latch prongs and retractor with the latch tailpiece as shown in Figure 19.48.

NO.	PART NAME	NO REQ'D
1	OUTSIDE KNOB AND SPINDLE ASS'Y	1
2	NYLON BUSHING	2
3	ROSE BUSHING	2
4	ROSE	2
5	OUTSIDE MOUNTING PLATE ASS'Y	1
6	INSIDE PLATE ASS'Y	1
7	MOUNTING SCREWS	2
8	INSIDE KNOB ASS'Y	1
9	SLEEVE & PIN ASS'Y	1
10	INSIDE KNOB COLLAR	1
11	WRENCH	1

Figure 19.38 Heavy-duty screwless trim. *(Alarm Lock Corporation)*

NO.	PART NAME	NO REQ'D
1	OUTSIDE HANDLE AND SPINDLE ASS'Y	1
2	NYLON BUSHING	2
3	ROSE	2
4	MOUNTING SCREWS	6
5	MOUNTING PLATE	2
6	INSIDE HANDLE	1
7	SET SCREW	1
8	WRENCH	1

OUTSIDE

INSIDE

Figure 19.39 Lever handle trim. *(Alarm Lock Corporation)*

NO.	PART NAME	NO REQ'D
1	OUTSIDE KNOB AND SPINDLE ASS'Y	1
2	NYLON BUSHING	2
3	OUTSIDE ESCUT.	1
4	INSIDE ESCUT.	1
5	MOUNTING SCREWS	8
6	INSIDE KNOB	1
7	SET SCREWS	2
8	CYLINDER RING	1
9	CYLINDER SPRING	1
10	MORTISE CYLINDER	1
11	KEYS	2

OUTSIDE

INSIDE

Figure 19.40 Medium-duty trim with the Glenwood escutcheon. (*Alarm Lock Corporation*)

NO.	PART NAME	NO REQ'D
1	OUTSIDE HANDLE AND SPINDLE ASS'Y	1
2	NYLON BUSHING	2
3	OUTSIDE ESCUT.	1
4	INSIDE ESCUT.	1
5	MOUNTING SCREWS	2
6	INSIDE HANDLE	1
7	SET SCREW	1
8	WRENCH	1
9	CYLINDER RING	1
10	CYLINDER SPRING	1
11	MORTISE CYLINDER	1
12	KEYS	2

OUTSIDE

INSIDE

Figure 19.41 Lever handle trim using the popular Utica escutcheon. (*Alarm Lock Corporation*)

Figure 19.42 Model 121 electrified mortise lock. (*Alarm Lock Corporation*)

Figure 19.43 Electrified key-in-knob lock installation, Step 1.

Figure 19.44 Electrified key-in-knob lock installation, Step 2.

Figure 19.45 Electrified key-in-knob lock installation, Step 3.

Figure 19.46 Electrified key-in-knob lock installation, Step 4.

BEFORE INSTALLATION MAKE SURE THAT CLEARANCE FOR
ANTIFRICTION LATCH IS CUT-OUT

Figure 19.47 Electrified key-in-knob lock installation, Step 5.

MAIN HOUSING

LATCH
TAILPIECE

RETRACTOR ENGAGES
LATCH TAILPIECE

LATCH PRONGS
ENGAGE MAIN
HOUSING

Figure 19.48 Electrified key-in-knob lock
installation, Step 5.

6. Secure the trim by replacing the rose and screwing it on to the threaded spindle with your fingers as far as it will go. Tighten the rose with the spanner wrench (Figure 19.49).

7. Replace the knob by lining up the keyway in the knob with the slot in the spindle. Push the knob straight on until the knob hits the knob catch. Depress the knob catch and push the knob straight on until the knob catch clicks into place. Test both knobs to make sure they are securely fastened to the spindle (Figure 19.50).

Figure 19.49 Electrified key-in-knob lock installation, Step 6.

Figure 19.50 Electrified key-in-knob lock installation, Step 7.

8. Install the strike by making a shallow mortise in the door jamb to align with the latch face and install the strike as in Figure 19.51.

Reversing the cylinder

If you must reverse the lock cylinder follow the procedure outlined in this section. First, insert the key into the cylinder and turn the key clockwise as far as it will go. At the same time, depress the knob catch retainer. When the retainer moves down, pull the knob off.

Figure 19.51 Electrified key-in-knob lock installation, Step 8.

Next, turn the knob 180°, and replace it as follows. Remove the key and push the knob on the spindle as far as it will go. Insert the key and turn it clockwise to its most extreme position while depressing the knob catch retainer. When the retainer moves down, push the knob onto the spindle, allowing the retainer to click into position.

Removing the cylinder

To remove the cylinder, first push the cylinder tailpiece sideways with your finger until the tailpiece unsnaps from the plug. Then, turn the knob over and hit the shank of the knob with a sharp blow on a flat surface. This releases the knob ferrule and permits the removal of the cylinder. Rekeying of the cylinder is conventional.

Electrical considerations

Now that you understand the specifics of installing the lock, consider its electrical components. The electrical power for the lock can either be concealed within the door by use of a current transfer unit (Figure 19.52) or left exposed on the interior side. The use of a current transfer unit is a unique means of bringing the electrical current from the lock side of the door to the door. The face and strike must be thoroughly matched and the door closed for power to flow through. With this in place, from a remote switch activation, power will flow through allowing the door to be opened.

The other method, leaving the wiring exposed, is not really what it seems. An armored door loop is used and it provides a much easier means for bringing electrical current from the hinge side of the door frame to the lock in the door. It consists of an 18-inch, flexible, armored cable that is attached to the frame on one end and just below the lock on the inside of the door. The lock wires can be passed through the door wall via a small hole and connected to the loop wire ends.

Electromagnetic Locks

While electric locks are available and used for many applications, there is also a need for electromagnetic locks. Such locks can be used in hospitals, convalescent homes, banks, universities, museums, libraries, factories, laboratories, penal institutions, and airports.

The electromagnetic lock, while operating at less than 12 Vdc, can exert a holding force of 1500 pounds. This is fantastic for a lock that consists of two components—the lock and its armature.

Figure 19.52 Current transfer unit. (*Alarm Lock Corporation*)

The lock mounts rigidly to the door frame and the armature mounts to the door (Figure 19.53). When the door is closed, the two make contact; on activation of the lock, the two are attracted to each other. The door, for all practical purposes, is bonded solidly to the frame.

Ten years ago, such a lock would have been priced out of reach, but today it is possible for almost any business or institution to afford one. As a locksmith, you must be prepared and familiar with this "unconventional" lock.

Security Engineering, the leader and major proponent of electromagnetic locks, lists these major features:

- A high holding (1500 pounds) effectively seals the door from both sides, keeping intruders out and/or valuables safely in.

- Conserves energy: The efficient low power consumption design (2 amps) ensures a long life and economical operating cost for the lock.

- High reliability: Recommended for high trafficked openings with extreme usage, where conventional locks, locks with moving parts, or electromagnetic locks with a lesser holding force may not be capable of withstanding the abuse.

- Fully fail safe: Locking and unlocking is positive and instantaneous, with no residual magnetism or moving parts that might stick, jam, bind, wear out, and/or need replacement.

- Extremely low maintenance: No maintenance or adjustments are required after initial installation.

- Exclusive mounting design: This permits rapid efficient installation in new or existing buildings and automatically compensates for normal door wear, sag, warping, or misalignment.

- Self-contained unit: It is completely factory wired with all options for the unit already built in, fully concealed, and tamper-resistant. There is only one external wire connection to make.

- Four operating voltages: 12 Vdc; standard 12 Vac; 24 Vdc; and 24 Vac.

Figure 19.53 Model 3900 electromagnetic lock. (*Security Engineering, Inc.*)

typical installation

110 VAC.

12 VDC

⑤

⑦⑥

WIRING
ACCESS
PLATE

①
③

②

④

1. Model #3900 Electric
   Lock
2. Electric Lock Armature
3. Door Status Switch
   Optional*
4. Local On/Off Station
   Controls

5. Power Supply
6. Remote On/Off Station
   Controls*
7. Hazard Sensing Safety
   Devices*

   *As System Requires

Figure 19.54   Model 3900 installation layout. (*Security Engineering, Inc.*)

Figure 19.54 illustrates a typical installation and the various features (standard and optional) that can come with a unit.

Series 3900 electromagnetic lock

It is important that, prior to installing an electromagnetic lock, you handle the equipment with care. Damaging the mating surfaces of the armature may reduce its locking efficiency. The electromagnet mounts rigidly to the door frame header. The armature mounts to the door with special hardware provided that allows it to pivot about its center to compensate for door wear and misalignment. All template use must take place with the door in its normal closed position.

The lock face is covered with a rust-inhibiting film. This should not be removed. If the film is accidentally removed, clean the lock face with a dry, soft cloth (do not

Figure 19.55 Door frame header determination tells you if a filler plate, angle bracket, or nothing will be required. (*Security Engineering, Inc.*)

touch the lock face with your hands) and reapply a rust inhibitor such as MI (manufactured by Starret) or LPS3 (manufactured by LPS Laboratories); these rust inhibitors are readily available in most hardware stores. Then, apply a light film of rust inhibitor generously to the armature face.

Mounting. Note the type of door frame header and install the filler place or angle bracket as required (Figure 19.55).

1. Fold the template (Figure 19.56) where indicated to form a 90° angle. For a swinging door, place the template against the door header and door opposite hinge side of the door jamb. For a pair of swinging doors, place the template against the door and door header at the center of the door opening. Transfer the hole locations to the door and door frame header.

2. Follow the template instructions for the specific hole sizes required. See Figure 19.56 to determine the proper armature mounting hole preparation. The hole is designated as the mark ⌽ in Figure 19.56.

Figure 19.56 Model 3900 lock template. (*Security Engineering, Inc.*)

3. Mount the armature to the door with the hardware provided as shown in the figures.

4. Install the mounting plate to the header with the five #10 screws.

5. Fasten the Model 3900 lock to the mounting plate with the two socket screws and compensate for any misalignment by adding or subtracting washers at the armature mounting screws.

6. Firmly tighten all the screws. Install the anti-tamper plugs into the holes over each socket head mounting screw. Use a soft hammer to avoid damaging the lock case.

Wiring instructions. In some instances, you may use a prefabricated power supply for the lock unit. What if you are not? To achieve the best possible electromagnet lock installation, note the following suggestions (both right and wrong ways are illustrated for each procedure).

- Proper placement of the on-off station control in the electrical circuit will ensure instantaneous door release and avoid any delay in complete unlocking (Figure 19.57).

- Care must be taken when adding any auxiliary controls or indicators to the electrical circuit. Improper placement may cause a slight delay in door release (Figure 19.58).

- Solid state on-off devices must be properly protected from the back voltage generated from the collapsing electromagnetic lock coil when the unit switched off. This may be accomplished by installing a metal oxide varistor (MOV) across the lock coil lead wires (Figure 19.59). The MOV is available from Security Engineering; see the address in Appendix B.

- Installation of the MOV is for arc suppression in ultra high usage installations to prevent switch contact deterioration. In any situation, though, the installation of a diode across the leads is not recommended (see Figure 19.58).

Series 2200 by Security Engineering

Another lock to consider is the surface-mounted electrically actuated lock, the Series 2200 from Security Engineering. This lock (Figure 19.60) was designed to offer

Figure 19.57 Model 3900 wiring.

Figure 19.58 Model 3900 wiring.

Figure 19.59 Model 3900 wiring.

Figure 19.60 Model 2200 series surface-mounted lock. (*Security Engineering, Inc.*)

the advantages of remote electric locking coupled with a versatility limited only by the user's imagination.

For basic operation, the lock mounts to a door frame or a door. When energized, it locks or unlocks the door when the solenoid-operated deadbolt projects from the unit and makes contact with a strike.

Standard features include the following:

- Fail-safe operation: If the power is interrupted, the unit automatically unlocks.

- Fail-secure operation: If the power is interrupted, the unit automatically unlocks.

- Easy installation: The unit is not handed and can thus be mounted on any plane.

- Versatility: It can be used with in-swinging, out-swinging, sliding, overhead, or roll-up doors.

- Maintenance free: Unit parts are ruggedly built and self-lubricating.

- Current draw: 0.5 amps at 24 volts.

- Low voltage: 24 Vac standard; 24 Vdc, 12 Vac, and 12 Vdc available.

- Long and silent operation: The unit has a continuous-duty solenoid that will function for millions of cycles.

- Compact size: 2 inches × 2 inches × 6¾ inches.

The unit has optional features including a choice of one or two signal lights; a built-in bolt position switch to indicate whether or not the bolt is extended or retracted for interfacing with automatic door equipment or signaling the lock status to a remote location unit. *Note:* The fail secure locks are not recommended where life safety may be compromised or where panic crash bar hardware is the only means of emergency egress.

Mounting.   Refer to Figure 19.61.

1. Remove the four screws from the unit; remove the housing cover.
2. Manually extend the solenoid bolt by pushing the BPS pin forward (if equipped) or inserting a wire into the rear of the solenoid.
3. Engage the extended solenoid bolt and strike and locate on the door and door frame as shown in the figure.
4. Using the mounting holes in the lock base and strike as templates, mount the solenoid lock and strike with the screws provided. (Several types of strike are available and can be used depending on the situation. See Figure 19.62.)
5. Check for the proper alignment and see that the lock is free, having no binding in operation.
6. Connect the electrical wiring per the system option or requirements.
7. Check the operation of the lock for any final adjustments; adjust as required.
8. Reinstall the housing cover and secure with screws.

Series 1300 mortise solenoid lock

A more compact but just as effective lock is the 1300 series mortise solenoid lock (Figure 19.63). This mortise solenoid-activated deadbolt is designed primarily for installation in new construction, with emphasis on tamper resistance and unit concealment.

On activation of the solenoid, the deadbolt projects ½ inch from the face plate or retracts ½ inch to flush with the face plate, depending on the solenoid selected and used.

Figure 19.61 Installation procedure for 2200 series lock. (*Security Engineering, Inc.*)

Figure 19.62 Strikes. (*Security Engineering, Inc.*)

Figure 19.63 Series 1300 mortise solenoid lock. (*Security Engineering, Inc.*)

The standard features of this lock unit are as follows:

- Fail safe and fail secure operation: The unit automatically unlocks when power is interrupted.
- Compact size: 1⅜ inch diameter × 4½ inches long.
- Voltage and current draw are the same as the 2200 series.

Electrical Keyless Locks

Simplex keyless security locks have also kept pace with electrically operated locks and latch units. The Simplex keyless entrance control is a wall-mounted unit control for electrical door locks (Figure 19.64) and has a built-in, adjustable, time delay feature not found in other, currently available units. Eliminating key control problems in high-risk areas, for home or business, is a distinct and great advantage.

The Simplex entrance control is an easy yet very secure installation. The unit is secured by two rugged draw bolts that pass completely through the wall. The bolts can be cut to any desired length to permit installation in walls from 4 inches to 11 inches thick (Figure 19.65). The adjustable timer delay of 15 to 20 seconds permits locating the control at any convenient position near or by the door.

Two styles of front mounting plates are included to meet varying installation needs. One is 3¼ inches wide × 9⅝ inches high; the smaller is 1¾ inches × 8½ inches. Both front mounting plates are made of stainless steel.

Security features for this unit include the following:

- A special visual shield for each entrance control unit to protect against unauthorized observation of the button sequence operation.

- If the control knob is forced by pliers or other tools, it will spin free without activating the electrical circuit.

- The electrical terminals and wire leads are protected against a short circuit attach by a heavy-duty metal back cover that completely encloses the entire electro-mechanical device inside the wall.

Figure 19.64 Wall-mounted electrical control unit for door lock or latch. (*Simplex Access Controls Corp.*)

Figure 19.65 Lock unit can be mounted on any thickness wall up to 11 inches. (*Simplex Access Controls Corp.*)

The Simplex/Unican Model 2000-15 is used by many businesses (Figure 19.66). It can be used with many leading brands of surface-mounted panic exit devices. Another popular unit is the Model TM pushbutton lock (Figure 19.67), which can be used in both businesses and homes.

The Memorilok system

A sophisticated digital access control system that does not require a console board or monitoring station is available. The Memorilok system has two basic units: an access control keyboard (Figure 19.68) and a program unit (Figure 19. 69). With these two units, 10,000 combinations are available to the customer. Now there is a third optional piece of equipment, the card reader (Figure 19.70), that provides a substantially higher degree of security integrity.

Figure 19.66 Unican 2000-15 unit used with an emergency exit panic bar. (*Simplex Access Controls Corp.*)

Figure 19.67 Model TM lock installed in an office cabinet. (*Simplex Access Controls Corp.*)

Figure 19.68 Pushbutton control unit.

Figure 19.69 Programmer unit.

Figure 19.70 Pushbutton unit with card reader for increased security.

The Memorilok system incorporates a solid state electronic memory for the storage of the selected code combination. The entry code combination is electronically stored. Unlike other access control systems, there are no wheels, dials, or other visual indicators to betray the code combination.

Standard unit features include:

Automatic address. A preset timer is activated when the first digit of the code combination is entered. The remainder of the code entry must be completed before the expiration of this time or the entry will be canceled and the programmer will activate the penalty time.

Automatic penalty.   A penalty time of approximately five seconds is incurred if an incorrect key is depressed during the code entry or if two or more keys are pressed simultaneously. This feature defeats random button manipulation.

Code bypass.   There is a provision within the access control unit for the connection of an external (remote) switch or pushbutton to operate the lock via the programmer unit without the use of the entry keyboard. This is normally sited within the protected access area and used for giving access to personnel or visitors who do not possess the code combination, but for whom access is warranted and authorized by the customer.

Memory protection.   The Memorilok keeps the code combination stored and undamaged in case of a main power failure. This is due to the constantly charged nickel cadmium battery contained within the unit.

Optional features include the following:

Duress/error alarm.   The duress/error alarm option is available as an integral part of the Memorilok unit and, if desired by the customer, must be ordered as original equipment. The duress alarm is activated by depressing a factory-fixed last digit (from 0 to 9) instead of the last digit of the selected code combination. If entry is made under duress, the first three digits of the programmed code combination plus the predetermined last digit—the Duress Digit—are entered on the keyboard. This automatically operates a set of Form C (Changeover) dry contacts that, in turn, triggers any desired alarm system: audible, silent, remote, or local.

The error alarm circuit will react on any number of selected unsuccessful attempts at entry. As soon as the predetermined number of incorrect attempts (from one to eight) have been made, the alarm is activated, warning that the system is being tampered with.

Flush mounting.   To flush mount the programmer, keyboard, or even the battery standby, an oversized front panel is used. The panels, 8 inches × 9 inches, are affixed to the standard front panels to conceal the rough opening around the Memorilok unit.

The optional card reader fits into the front of the keyboard panel. It provides higher levels of security integrity because the coded card *must* be used by the individual seeking entry. It must be entered into the Memorilok before the code combination will be accepted by the electronic memory circuit.

Wiring instructions.   Wiring details are in Figure 19.71. The following notes apply, as referenced in the figure:

- These connections are not used unless the remote entry/exit is required. Use the open switch.
- All contacts are shown (NO and NC) without the lock being energized and without being in the alarmed state.

- On installations where there is a long wire run between the programmer and the lock or other device being operated, carefully compute the minimum requirements of the lock or device and the voltage drop which will occur in the wire.

- Code bypass switch requires a momentary switch with normally open (NO) switch contacts.

- The alarm Reset Switch (R-S) requires a momentary switch with normal switch contacts. Thus should be wired between terminal #2 of the exit and terminal R-S indicated on the diagram as "blacked out block" on the programmer wiring hookup. Use this only for duress/error entry.

Programming the Memorilok. To program the Memorilok, turn the programmer key to the *on* position. Enter the desired four-digit code combination on the keyboard. A green LED indicator light glows to indicate that the code is properly programmed into the memory system. Return the programmer key to the off position. Reinsert the code to test the unit for proper operation. Now the green LED light indicates acceptance of the code combination and supply of power to the lock or other device used.

The adjustable time delay for the Memorilok affects the lock open time (time delay for the unit) and it can be varied from one to ten seconds by turning an adjusting screw.

Unit specifications are in Figure 19.72. Specification installation details, template, and other specific information concerning conductor wire are provided with the unit from the factory.

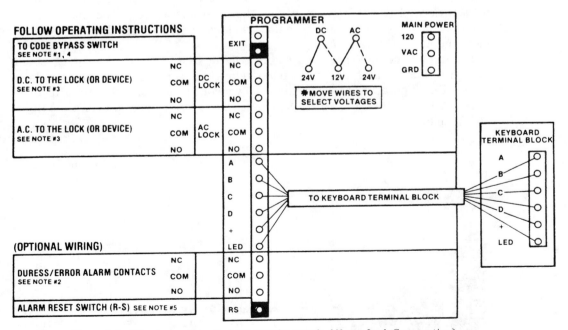

Figure 19.71 Memorilok wiring instructions (refer to text for notes). (*Alarm Lock Corporation*)

SPECIFICATIONS:

Dimensions:	
Keyboard Unit	5 ⁷⁄₁₆" (128 mm) Wide 6 ⁷⁄₁₆" (153 mm) High 2 ⅝" (66 mm) Projection
Programmer Unit	Same as Keyboard
Shipping Weight:	Complete Memorilok - 6 pounds Battery Standby Unit - 6 pounds
Main Power Input:	120 VAC or 240 VAC 50/60 HZ. Specify voltage required.
*Output to Lock:	12 or 24 VAC or VDC at 1 Amp. Voltage is user selectable.
Program Key Switch:	Contacts are closed in the "ON" position for code programming.
Wiring Information:	Includes ten feet of six conductor 22 gauge wire.

Figure 19.72   Memorilok 400 unit specifications. (*Alarm Lock Corporation*)

Key-Actuated Switches

The full-service locksmith should have available for customers, especially those in business, a variety of key-actuated switches. Figure 19.73 shows a Micro Kaba rotary switch. This miniature switch is operated by the 12-millimeter diameter cylinder, which has over 10,000 different possible key combinations. Four key entry variations are available, each having up to four switch elements (Figure 19.74).

Two-pole switches (Figure 19.75) are also available. They provide for 90° movement with two key withdrawal positions, thus allowing the electronic unit to be held in either the on or off position. The unit can take up to 250 Vac, at 4 amps. Unit dimensions are in Figure 19.76.

The surface-mounted remote control stations (KS RCS/1 and /2) are die-cast aluminum with a stainless steel front plate, tempered against opening, and a red and green LED (Figure 19.77). The waterproof version ensures unit use during all types of weather conditions. Figure 19.78 provides basic wiring information for the RCS/1 and /2 units.

Figure 19.73   Micro Kaba rotary switch. (*Lori Corp.*)

KEY ENTRY		No. of Switch elements
	Spring Return	1 2 3 4
	One Key Entry 9h	1 2 3 4
	One Key Entry 6h	1 2 3 4
	Two Key Entry	1 2 3 4

Figure 19.74 Four different key positions provide an opportunity for different control features. (*Lori Corp.*)

Figure 19.75 Two-pole switch. (*Lori Corp.*)

Figure 19.76 Specifications for the two-pole switch. (*Lori Corp.*)

Figure 19.77 Surface-mounted remote control station with a waterproof key cover. (*Lori Corp.*)

1 - 2   **Tamper Loop**
3 - 4   **12 Vdc LED "on"**
6 7 8   **ON/OFF Changeover Switch. KS64**

WIRING DIAGRAM FOR REMOTE CONTROL STATIONS KS - RCS/1

1 - 2   **Tamper Loop**
3 - 4   **12 Vdc LED "on"**
6 7 8   **ON/OFF Changeover Switch KS64**

WIRING DIAGRAM FOR REMOTE CONTROL STATIONS KS - RCS/2

Figure 19.78 RCS/1 and RCS/2 wiring diagrams. (*Lori Corp.*)

Another remote control station with an anti-tamper switch is in Figure 19.79. Like the other units, it uses the Mini-Kaba or the Kaba 20 key/switch. A remote control station mounting box (Figure 19.80) is available and used for mounting the unit. Mounting box dimensions are in Figure 19.81.

Remote control switches are UL approved. Electrical specifications are in Figure 19.82.

Figure 19.79 The narrow remote control switch allows for many different installation applications. (*Lori Corp.*)

Figure 19.80 Remote control station housing mounting box. (*Lori Corp.*)

Cut out dimensions 83mm x 32mm

Figure 19.81 Mounting box specification data. (*Lori Corp.*)

Rated Voltage (V)	Non-Inductive Load		Inductive Load	
	Resistive (A)	Lamp (A)	Inductive (A)	Motor (A)
AC 110 to 125	5	1.5	3	2.5
220 to 250	5	1.0	2	1.5
DC 6 to 8	5	2	5	3
110 to 125	0.4	0.05	0.4	0.05
220 to 250	0.2	0.03	0.2	0.03

Figure 19.82 Electrical specifications for the remote control station. (*Lori Corp.*)

The Business of Locksmithing

When you consider starting your own locksmithing business, keep in mind the following factors.

Police clearance. Many jurisdictions require licensing and fingerprinting.

Professional qualifications. Some cities insist on tests to determine professional competence.

Basic tools and equipment. Ensure that you have all of the equipment you need to tackle any locksmithing job.

Supply stock. It's possible to work out special arrangements with some local supply houses.

Financing. Setting up shop involves an initial cash outlay.

Bookkeeping. You must keep records of all transactions.

Vehicles. A vehicle is necessary for out-of-shop calls.

Advertising. Sometimes it takes more than just competent locksmithing to bring in business. Good advertising usually includes signs, business cards, window displays, display racks, display materials, etc.

Literature. You need references and materials including locksmith supply house catalogs, manufacturers' literature, reference books, business forms, etc.

Contacts. It pays to know other locksmiths, factory representatives, etc. They can help you with locksmithing problems from time to time.

Locksmithing is a profession involving skill, competence, and public trust. It is a career that can be highly rewarding. This chapter describes the background necessary to use the basics of what you have already learned and apply it in a business.

The Occupational Outlook Handbook indicates that in 1972 there were approximately 8000 full-time locksmiths in the United States. Many of these 8000 individuals operated their own locksmithing business, but others were in a shop which employed one to three people. The situation is very much the same today, meaning that in this country, the demand for competent locksmiths is overwhelming.

Fifty to 75 million keys are lost and replaced each year. Every household, automobile, and business has two to five locks. In a medium-sized city of 75,000 to 100,000 people, there are easily more than 250,000 locks. This means there are literally billions of locks and keys in this country—plenty of business for a good locksmith. Statistics show that locksmiths make good money, too.

Business Considerations

If you're thinking about launching a career in locksmithing, there are several things you should consider.

Selecting a business site

Selecting your business site is a major decision and one that will affect your business throughout its infancy.

In determining a location for your business, consider the types of businesses you will be near and what effect they will have on your customers. Shopping malls are ideal for locksmithing shops because of the walk-in business generated by people who need one or two keys duplicated. If you undertake home servicing, you will want a location that enables you to get to your customers quickly.

The type of building must be considered. It would be unwise to obtain a building that is in need of massive repairs. The building should be large enough for your needs. Thought should be given to the building's potential for future expansion.

Demographic and economic factors of the surrounding area should be considered too. Make a study of population density, population growth, projected economic growth, the tax structure, etc. Examine any potential competition with other locksmiths.

Shop layout

Window displays are a good way to attract customers. The display itself should be simple and effective, neat, and clean. Window displays cannot be cluttered; you must be discriminating and orderly in developing them. The display should tell the viewer something interesting and appealing about your products or services.

The front counter area should be carefully organized; this is where sales are made. Ensure that the key machines are near the counter—visible from the street. Spread out your keys on a keyboard; such a display stays in your customers' minds longer than a pile of keys on a table. Code books usually are kept near the key machines.

The workbench usually should be out of the customers' sight, preferably in the back room. If it is out front, it should be clean at all times.

Lighting of the work and customer service areas is most important. It should be adequate but not overlighted. The lights should be of the non-glare variety. The work area can have additional small lights placed several feet above the bench.

Merchandising through advertising

Advertising takes many forms, from radio, television, and handbills to putting your name on each key duplicated. There are no hard and fast rules. Good advertising is advertising that gets results.

The selection of the proper advertising can be important for a beginning locksmith. Consider the population; do they listen to the radio, watch television, or read the paper more than in other areas, or do they look for gimmicks? What are the costs of the various advertising media? You can approach the advertising staffs of the local media to determine these answers.

Once you complete an advertising scheme, evaluate the results. Did the advertising increase business? Was the advertising worth the investment?

Ultimately, the locksmith herself is the most effective advertisement. Responsible locksmithing, a neat appearance, and a little courtesy go a long way. In spite of all the sales talk and advertising hoopla, the customer is only interested in one thing—quality service.

Whether you work out of your home, your vehicle, or a formal locksmithing shop, you need to have some form of advertising at your shop. A sign outside that merely states "Locksmith" isn't really enough. Whenever possible, you should have and use small stickers that can be affixed to windows, vehicles, etc; they can increase your business.

Figure 20.1 shows a variety of such aids, including decals, window banners, and a sign. Whenever possible, you should use these types of aids. They promote business and tell your potential customers exactly what services you can or cannot provide. Signs also save you time in answering general inquiries from walk-ins off the street who are not sure if you can provide a specific service, such as cutting a circular (Ace) key. Having a sign indicating that you duplicate these keys can increase business.

A hanging sign (Figure 20.2) with a slightly different slant to it (Keys Cut While You Watch) is another visual aid to customers. By using properly aligned key machines, you can ensure that the phrase "guaranteed fit" is accurate.

Decals for your office or on the service van also aid in promoting your business and security in general (Figure 20.3).

Selecting key blanks

It is important to carry a variety of key blanks to meet the continuing needs of your customers. More important, you must determine which blanks are most popular in your area. Purchasing one or two boxes of every key blank that you might use during your career in locksmithing is ridiculous and extremely costly. Purchasing such items randomly shows poor business judgment. Large companies that supply all kinds of locks and security devices will, naturally, have the widest variety of key

Figure 20.1 Various sales aids used in locksmith shops. At upper left is a key sign; others are window banners. (*Dominion Lock Co.*)

Figure 20.2 Advertising sign to hang in window of locksmithing shop. (*Dominion Lock Co.*)

Figure 20.3 Decals for use in office or on side of van. (*Dominion Lock Co.*)

blanks available for your purchase and use. That assortment doesn't mean that you really need such an extensive variety of key blanks. In almost every case, you do not.

To decide what you should have, examine the key determination process from the viewpoint of a key blank manufacturer. The following information comes from Herbert Stein, President of Star Key & Lock Manufacturing, Co., Inc., of New York. For many years, this company has been in the business of carefully foreseeing the key blank needs of locksmiths.

The first step is to consult a number of the largest competitors in different parts of the country for their opinions of the relative popularity of a particular blank and the anticipated annual demand. (Some blanks may be extremely popular in certain areas of the country, but very slow in other parts.) Then, consider such factors as whether the blank's lock is made by a major company or not and whether it is a big or small volume lock (for example, if the blank is for an automobile, is it for the major line of one of the top domestic or foreign automobile makers, or is it just for one of the small volume imported cars?). Naturally, if a deluge of requests for a specific type of blank comes in from across the nation, it will gain priority on the production list.

Knowing how a key blank is selected for production helps, but how do you find out what the most common blanks are in your area? Here is where your regional representative of the manufacturer, or direct contact with the factory, can help. Knowing what keys are most used in your area is part of their job. Through various national and regional trends, regional and area market studies, and previous sales to your area, the manufacturer can determine with great accuracy what the most common and popularly requested blanks will be for any given part of the country.

While manufacturers keep such records, you might feel that you are imposing on them with a "trifling request." Not so, but even then, some locksmiths just look through a catalog, especially when starting out in a business, and contact a local locksmith distributor/supply house in the area. The supply house can assist you, but many sales of key blanks don't go through the supply house; they are purchases from other area distributors or directly from the manufacturer. The supply house, while in touch with your area, isn't always as fully cognizant as the manufacturer in this regard.

At the supply house you will probably be shown one or several catalogs with a myriad of key blanks with various ones indicated that you should have on your shelf. These may or may not meet your requirements. Again, Star Key & Lock has thought of this; their catalog lists the most commonly used key blanks for just about every situation. This is where it is to your advantage to deal directly with the manufacturer. By purchasing assortments of different types, you reduce your initial key blank costs, reduce your stockroom inventory, have more space for other products, and know exactly what your customers will expect. In addition, you don't have several hundred or more blanks to look through for any given type, but perhaps only a hundred and fifty or so.

By purchasing an assortment, you have the advantage of getting the keys in display mounts that provide a subtle buying hint to the customers who walk in to your shop. These displays hold the most popular key blanks (usually between 90 and 150 different) that will meet just about all of your needs. The few oddball keys that come in for duplication are also the ones that give every locksmith a headache, because it may be only once every two to five years that a key of that specific type is needed.

A little footwork at your end helps you assess the needs of your shop. Check out the local general supply houses and see what blanks are the most popular with them. Ask them about the most commonly purchased locksets in your area and get to know these locks, the different keyways possible and available for them, and also what the big sellers were within the past ten years—not just the big sellers in today's market.

From all of this information, you should have learned:

- The most popular key blanks in your area
- The volume of blanks used in the past several years for different types of locks
- The top dozen locks, by manufacturer, selling currently in your area
- The top dozen locks, again by manufacturer, that were big sellers during the past decade
- The various keyways necessary for these locksets

This knowledge (coupled with the size of your shop and projected business) gives you a very good idea of which key blanks will be the most in demand, which ones will require a medium stock, and which ones you have to consider only once every two or three years. The minimal stock blanks will last for a long time, but the more popular ones may need frequent restocking—possibly every month.

After you complete your research, you can properly select the blanks you feel are necessary. Remember that it is probably best to consider the general assortment of key blanks when first starting out. Get the catalogs that provide these blanks and consider their assortments. Later on, as your business expands, you can move on to include other key blank manufacturers for, less common blanks that you see a need to stock.

After completing your key blank research, you should also know the types of locks you require, as well as exactly what types of key reference catalogs for key codes your shop should carry. You can start listing the specific cylinders, pin sizes, and dimensions required, specific repair parts kits, key machines that may be necessary, and some other specifics that will be necessary to perform as a locksmith.

"Instant" blank identification. The professional locksmith knows his or her key blanks and doesn't need to constantly refer to charts and catalogs for this information. In the beginning, every locksmith has trouble identifying one key among hundreds that may be on the shop wall.

Star Key & Lock has come up with a nifty idea that is extremely helpful (and something that I wish that all manufacturers of key blanks would do): they put an "instant" identity on the key blank. This new feature has put Star far out in front of other manufacturers. What it amounts to is stamping the make of the lock on the key blanks in addition to the comparative numbers. This achieves immediate identification of the key blank by the locksmith, saving time and trouble, and is a valuable learning tool for student and apprentice locksmiths. The make of lock stamped directly on the key also helps convince the customer that he is receiving the correct key for his lock, which aids in establishing the locksmith's credibility.

Embossing. Embossing is an added feature that you can supply to customers who require something special or an identifying logo. This means imprinting the bow of the key blank with anything from a fancy logo design to several lines of advertising copy. Imprinting can be done on any number of the brass keys available from Star. This is not a common or standard service, though some other manufacturers offer it as well.

Special embossing usually consists of three lines or less of standard type copy. This is based on a maximum of 14 figures on the first line, 12 on the second line, and 9 on the third line. The manufacturer must make back the costs of embossing; as a minimum, consider an order of 1000 embossed keys of assorted key blank numbers in increments of at least 50 pieces per number ordered.

Embossing is an excellent way to advertise your business. It costs less than average advertising and still gets your message across to the public that needs and uses your products and services (Figure 20.4). The chief advantages of special embossing are excellent advertising, increased locksmith shop prestige, and indisputable proof that the keys were made by the name embossed thereon.

Key materials. Most keys in use today are made of brass. This is important. All-brass key blanks are best for impressing because they have structure which shows distinct impression marks. The blanks cut very easily on key machines or by hand filing, but don't believe the myth that brass is soft; it's not true. The all-brass key retains maximum strength (77,000 p.s.i.); the special brass alloy which is used by Star Key & Locks (and other manufacturers) gives the best combination of desirable properties of impression-sensitive structure, free-cutting duplication, and the maximum strength necessary for a key blank.

Both as a locksmith and as an individual who requires keys in your own day-to-day movement (home, auto, office), you will see through experience that certain key blanks are better than others. Use the products that are of the best quality, not those that cost the least

Key cutting machines

Figure 20.5 shows a key machine for the office or for use on the job in remote locations. This compact, highly dependable, cylinder key cutter, the "Companion," will

Figure 20.4 Excellent advertising increases the locksmith's prestige and provides evidence that the keys were made by you. (*Star Key & Lock Co.*)

Figure 20.5 The "Companion" model portable key machine operates off 110 Vac or 12 Vdc current. (*Dominion Lock Co.*)

go wherever your locksmithing work may be. It has dual voltage (110 Vac or 12 Vdc) capability with no extra wiring necessary. It comes in a sturdy carrying case and weighs only 25 pounds.

The machine has four-way jaws to grip any cylinder key; no special adapters are required. The precision alignment of the cutter ensures a correctly duplicated key everytime. (Of course, check the cutter alignment periodically to ensure that the precision alignment is true. Rough handling and occasional bumps may throw the alignment out slightly. Be safe, be sure, and cut a true duplicate key every time.)

The key display and machine is a visual selling aid in your locksmith shop. Usually located near the counter or cash register, it promotes sales subtly by showing your customers that you have the commonly used keys in stock and can duplicate a key for them while they wait. In the case of the display in Figure 20.6, it is important that you know exactly what is available. This particular key assortment and cutting machine is a self-standing, revolving counter rack and covers both U.S. and foreign automotive keys, in addition to keys for business and residential locks,

The jaws on this key machine can be rotated to present four different gripping surfaces to cut four different types of keys without using any adapters. The jaws simply lift up and rotate, eliminating the fumbling necessary with adapters for the key machine. The *standard* station of the cutter jaws holds regular cylinder keys, while the *wide* jaws grip Ford double-sided keys. The *A* station jaws hold the Schlage wafer key (SC 6) and the *W* station jaws are for the Schlage water key (SC22). Other keys, such as Chicago and double-sided Datsun keys, can be cut with the *A* station jaws.

Key machine cutters.   After continued use, the cutters for your key machine(s) will become dull. This, in turn, means that each key you cut will not have the precision cuts required. Any such key may operate in the lock roughly or not at all. To alleviate this, you should have at least one replacement cutter of each type you normally use on hand.

The basic cutters are shown in Figure 20.7 and include the file, slotting, milling, and side milling slotter. The cutter specifications and key type uses are as follows:

Material	Use	Cutter Type
File	High carbon steel	Cylinder keys
Slotting	High speed steel	Flat keys with 0.045-inch cut*
Milling	Tungsten-chrome alloy steel	Cylinder keys
Side Milling	High-speed steel	Cylinder and flat keys (0.045-inch slotter cut)

*Two different slotting cutters are available with different thicknesses for 0.030-inch and 0.045-inch cuts on flat steel keys.

Figure 20.6 Key machine with common key blanks for quick duplication. (*Taylor Lock Company*)

Figure 20.7 Basic replacement key machine cutters, as well as a file, slotting, milling, and a side-milling slotter. (*Taylor Lock Company*)

Displays

The color key carousel is shown in Figure 20.8. Because colored keys are becoming more popular, some customers will prefer to have another color key when having their key duplicated. Color keys assist the customer in rapid identification of specific keys. In some instances, they will not accept a key that is not colored.

A key comparison board allows you to rapidly determine which blanks, by manufacturer, are the same in their component makeup (i.e., the specifics of the key identification). A key comparison board is a great advantage to the locksmith and even more so for the apprentice or locksmith trainee.

In the case of these two items (as with others that you find in various catalogs), the maker isn't just hyping a product in hope that you will buy it; the manufacturer genuinely wants to provide you, a locksmith, with items that will assist you in the locksmithing trade. By helping you to do your job a little better, easier, faster, and more professionally, and by making available the supplementary business items required, the manufacturer is helping everyone: you, your customer, and himself. When you have problems, questions, or need assistance, the manufacturer or his representative is there to help you. Likewise, the manufacturer is looking ahead to your varying needs and has a variety of items to help you to help yourself.

Another example of a manufacturer's concern for assisting the locksmithing business is shown in Figure 20.9. Lock displays like this unit are available for setting up and demonstrating specific products within your shop. The customer is interested in a specific type of product—in this case, a lock. With a variety of these displays available, the customer can try out each one and, as the choice of locks narrows, you provide specifics concerning the remaining locks. In this particular case, you might point out the very salient and positive features of the Ultra 800, the Medeco cylinder, the strength and positive locking action of the mechanism, and other specifics.

Figure 20.8 Color key carousel.
(*Star Key & Lock Co.*)

Figure 20.9 Manufacturer-prepared lock display for customer consideration. Every shop should have a variety of such displays for customers to view and touch. (*M.A.G. Engineering and Mfg., Inc.*)

For narrow stile door installation, Adams Rite has created a specific installation kit. The advantages of this kit should be obvious: extra parts and detailed information for each stile installed. For the locksmith who plans continued installation, repair, and/or replacement of lock units in narrow stile doors, this kit is a must. It allows for a smoother, more rapid installation of numerous units, saving a tremendous amount of time (Figure 20.10).

The Locksmith and the Law

As a locksmith, you must have a better understanding of some laws than most people. When setting up your business, it is prudent to consult your attorney regarding all laws that concern your profession. In many jurisdictions there are laws covering licensing, control of locksmithing tools, and the registration of code books. Some local laws regulate the conduct of certain locksmithing business practices, such as duplicating master keys, making bank deposit box keys, opening automobiles, etc.

When first entering the business, aside from consulting with your lawyer and possibly the police, you should also contact your area locksmithing association, if one

Figure 20.10 Installation kit for the Adams Rite narrow stile door units. (*Adams Rite Mfg. Co.*)

exists. You can also obtain information from national locksmithing organizations through their newsletters and publications. These are excellent sources of information about your legal responsibilities.

Your legal responsibilities demand that you be very careful about the jobs you accept. You must be certain that you are not breaking the law by complying with a customer's wishes. Authorization statements from supervisors, such as from a bank or post office, for duplicating a key should always be double-checked with the main offices. Verification of such written statements is a must; they should also be filed with the job order.

When jobs of this type arise, the following information should be included in your files:

- Name of the person who brought the job in
- Identification (Social Security card or driver's license)
- Address of the person who brought in the job
- Business telephone (call to double-check)
- Name of the firm
- Business address
- Type of service performed
- Type of payment (if a check, it should be a business check, not a personal check)
- Automobile make, model, serial number, license number, state in which it is registered

You may also have forms printed up for your customer to fill in. You can use a single form for all of the situations you run into. Keep the form with the work order.

In some states and cities, laws require that you take an examination administered by your locksmithing peers to show that you are a competent locksmith who meets professional standards. Besides this, the police may check your reputation, qualifications, background, and previous employers to ascertain that you are of good moral character.

Various laws have been enacted both to protect the public from unscrupulous individuals posing as locksmiths and to protect the locksmithing trade from such individuals. The following is extracted from the Los Angeles Code, Ordinances No. 83,128, as an example of a local law regulating the locksmithing profession:

Sec 27.11 Locksmith—Regulating Applicable To

A. Definitions

"Locksmith" shall mean any person whose trade or occupation, in whole or in part, is the making or fashioning of keys for locks, or similar devices, or who constructs, reconstructs or repairs or adjusts locks, or who opens or closes locks for others by mechanical means other than with the regular keys furnished for the purpose by the manufacturers of the locks.

B. Trade of locksmith—permit required

No person shall engage in the business of locksmith, or practice or follow the trade or occupation of locksmith, without a permit therefore from the Board of Police Commissioners.

C. Permit—Application

Such permit shall be issued only on the verified application of the individual seeking the permit. The application shall be on a form prescribed by the Board and shall set forth the proposed location of the applicant's place of business, the names and address of five character references, and such other things as the Board may require to determine the character, honesty, and trustworthiness of the applicant. Specimen fingerprints of the applicant shall be furnished with the application.

D. Permits—fees—expiration

Each application for a permit shall be accompanied by a fee of $10 and each application for the annual renewal thereof, by a fee of $5. Each permit, unless sooner revoked, shall expire on December 31st, following the date of issuance. Each permit shall bear a serial number.

E. Permits—issuance and denial

The Board shall cause an investigation to be made on each application, and if the Board finds that the applicant's reputation for honesty is good, that he has not used his skill or knowledge as a locksmith to commit or aid in the commission of burglaries, larcenies, thefts, or other crimes, that he intends honestly and fairly to practice the trade of locksmith in a lawful manner, and that he has not been convicted of a felony, then the permit shall issue. Otherwise it shall be denied.

F. Permittee—must keep record

Each permittee must keep a book, which shall be open to inspection by any police officer at all times, in which the following must be entered:

1. The name and address of every person for whom a key is made by code or number.

    2. The name and address of every person for whom a locked automobile, building, structure, house, or store, whether vacant or occupied, is opened, or a key fitted thereto.

G. Keys to be stamped

It shall be unlawful for any locksmith to fail to stamp the serial number of his permit on any key made, repaired, sold, or given away by him.

H. Signs to be displayed

Every locksmith shall display in a conspicuous manner in the place where he is carrying on such business, trade, or occupation, a sign of a style, size, and color to be prescribed by said Board, reading, "Licensed Locksmith," together with the official permit number.

I. Permits—revocation

The Board, on proceedings had as prescribed in Section 22.02, may revoke or suspend any permit issued hereunder on any of the following grounds:

1. Misrepresentation in obtaining such permit.
2. Violation of any provision of this section.
3. That the permittee has committed or aided in the commission of or in the preparation for the commission of any crime by the use of his skill or knowledge as a locksmith or by using or letting the use of his tools, equipment, facilities, or supplies.

(Note: To learn more about locksmithing laws, ethics, and business matters, refer to my book Professional Locksmithing Techniques, 2nd Edition, TAB Books/McGraw-Hill.)

Key Duplicating Machines

One of the most important investments you'll make in locksmithing equipment is a key duplicating machine. A wide variety of them are available. Prices range from a few hundred dollars to several thousand dollars. Figure 21.1 shows a few key duplicating machines.

A wide difference in price doesn't necessarily indicate a substantial difference in functions and quality between key duplicating machines. This chapter will help you better understand how the machines work and how to find the best one you can afford.

A key duplicating machine consists of four basic parts:

- Two vises move in unison. One vise holds the key being duplicated; the other holds the key blank.
- A key guide traces the profile of the key.
- A cutter wheel cuts the key blank in accordance with the profile traced by the key guide.

Most inexpensive key duplicating machines are designed to duplicate only cylinder keys, but a few inexpensive models can also duplicate flat keys. More sophisticated models duplicate other types of keys. The most expensive ones can duplicate bit keys, tubular lock keys, angle keys (such as Medeco), and dimple keys. You can save money by purchasing a machine that duplicates only the types of keys you're most interested in duplicating.

Critical Design Factors

When evaluating a key duplicating machine, look carefully at the pivot mechanism. The key should meet the cutter wheel squarely on a dead parallel with the axis of the wheel. A slight angle is enough to upset the dimension of the duplicated key. (Figure 21.2.)

Figure 21.1 Key cutting machines. (*Ilco Unican Corp.*)

Figure 21.2 A key machine should be constructed so the key is almost parallel to the cutter. Along pivot (shown on the right) reduces the angle.

Some machines don't have pivots and arrange matters so the vises move laterally into the cutter. As long as the bearings are true, this ensures that the duplicated key is a faithful copy of the original.

The jaws of the vises must be carefully engineered to ensure that both the key and key blank do not shift during the duplication process and that the key and key blank are held squarely against the cutter.

Cutters

There are three basic types of cutters. *Filing* and *milling* cutters are both used to duplicate cylinder keys. A *slotter* cutter is used for square-ended keys (such as bit keys and flat keys). Figure 21.3 shows some examples.

The diameter of the cutter is important. Large diameter cutters leave a very slight concave on the newly cut key. Smaller cutters make a deeper concave. The most useful key duplicating machines allow you to use various sizes and types of cutters.

Key guide

The key guide should be checked regularly. In the absence of manufacturers' instructions for doing so, check the guide by mounting two identical key blanks in the vises; then, lift the vises up to the key guide and cutter wheel. While keeping the vises in place, slowly rotate the cutter wheel. The cutter wheel should barely scrape one of the key blanks. If the cutter wheel doesn't touch a key blank, or if it digs deeply into a key blank, the machine is out of alignment and needs to be adjusted. Figure 21.4 illustrates how to check a key guide.

Framon's DBM-1 Flat Key Machine

Framon Manufacturing Company's DBM-1 is a high-quality, flat-key duplicating machine. It's designed to duplicate a wide variety of flat and corrugated keys (Figure 21.5). By understanding how it works, you'll be able to operate many similar machines.

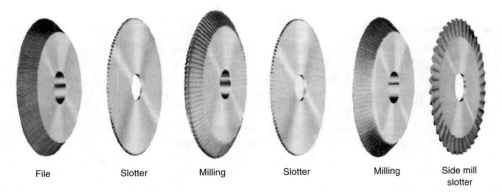

| File | Slotter | Milling | Slotter | Milling | Side mill slotter |

Figure 21.3 Various cutter wheels.

Figure 21.4 The cutter wheel should barely touch the key blank (left). (*Keynote Engineering*)

Figure 21.5 Framon's DBM-1 Flat Key Machine. (*Framon Manufacturing Corp.*)

Cutting procedure

All keys should be set from the tip for spacing. Insert the pattern key in the right hand vise with the tip of the blank protruding slightly beyond the left side of the vise (Figure 21.6). This position allows cutting of the tip guide on the blank if the blank tip is slightly different than the pattern key. In this position, the tip can be cut without the cutting vise.

Push the guide shaft to the rear and lock it into this position by tightening the locking knob (Figure 21.7). This relieves spring pressure so the tip setting is easier. Lift the yoke and set the tip of the pattern key against the right side of the guide. While holding this position, insert the blank and set the tip of the blank against the right side of the cutter (Figure 21.8). This procedure assures proper spacing.

Release the guide by loosening the locking knob; the key is now ready to be cut.

See the cut in the pattern key against the guide and lift the yoke into the cutter to make the cut (Figure 21.9); lower the yoke and repeat for the next cut. Follow this procedure until all cuts (including the throat cut) are made and the key is complete. All cuts should be made with a straight-in motion. This ensures clean, square cuts.

Notice that there is no side play in the guide assembly, so all cuts on duplicate keys will be the same width as pattern keys.

Figure 21.6 Set key from tip for spacing. (*Framon Manufacturing Corp.*)

Figure 21.7 Tighten locking knob to lock guide shaft into position. (*Framon Manufacturing Corp.*)

Figure 21.8 Set tip of blank against right-hand side of cutter. (*Framon Manufacturing Corp.*)

Figure 21.9 Set cut in pattern key against guide. (*Framon Manufacturing Corp.*)

Adjustments

With tip setting for spacing, there is no problem with improper spacing.

Depth setting can be checked by using two blanks that are the same. Check the depth by drawing the blank against the cutter guide. If depth adjustments are needed, simply loosen the set screw on the depth ring and adjust the ring until the cutter barely touches the blank. To make cuts deeper, rotate the depth ring counterclockwise. To make cuts shallower, rotate the depth ring clockwise.

Tighten the set screw after the adjustment is made, but do not tighten it too far.

Another way to check depth is to make one cut on the duplicate key; check the depth cut on both the pattern key and duplicate. As an example, if the cut on the duplicate key is 0.003 deeper than cut on the pattern key, loosen the set screw on depth ring, rotate the ring 0.003 clockwise and tighten the set screw. *Note:* Calibrations on the depth ring are in increments of 0.001.

Check the guide setting; the guide must be set to the same width of the cutter used. The DBM-1 is supplied with one 0.045-width cutter. This is the best width for general work. Cutter widths of 0.035, 0.055, 0.066, and 0.088 are available if needed. Cutter width of 0.100 (LeFebure) can be obtained by using a 0.045 and a 0.055 at the same time. All of these cutters are solid carbide.

To set the guide, simply loosen the cap screw (DBM-42). Rotate the guide to cutter width and tighten the cap screw. The Detent screw in the guide shaft will align the guide. No adjustment is required when changing guide settings.

Maintenance

Lubricate the yoke rod (DBM-09), guide shaft (DBM-09), and vise studs (DBM-18) sparingly using very fine oil. (*Do not use motor oil.*)

Wipe off all excess oil. To lubricate the guide shaft, unscrew the locking knob (DBM-21) and put one or two drops of oil in the opening. Replace the knob.

Other than these parts, cleanliness is the best maintenance.

Ilco's KD50A

Ilco Unican Corporations' KD50A is a high-quality machine for duplicating cylinder keys. By understanding how it works, you'll know the basics of operating most other cylinder key duplicating machines. The following information about the KD50A came from an instruction manual published by the manufacturer.

General operating sequence

The KD50A has a constant power switch that must be turned on. However, the machine motor will not operate until activated by the carriage assembly (Figure 21.10).

After both key and blank are properly clamped and aligned, pull down on the carriage handle. Use your thumb to depress the carriage release knob; the key setting gauge will automatically spring away. Spring tension will raise the carriage and the motor will automatically start.

Move the lever handle sideways so that the original key touches the key guide in an area between the shoulder and the first cut. Do not let the shoulders touch either the key guide or cutter wheel. Using the lever handle, slide the carriage left and then right to complete the cutting operation. Lower the carriage until it locks into the original position, which will automatically stop the motor and the cutter. Remove the new key and deburr with the brush; do not overbrush or run the key into belts.

NOTE: On/Off power switch (KD50A-15) is not shown, but is visible on the left side of the machine.

Figure 21.10 Ilco Unican's KD50A Key Duplicating Machine. (*Ilco Unican Corp.*)

Adjusting for proper depth of cut

For safety, remove the wire plug from its electrical socket. Clamp the two service bars into the vise jaws as shown in Figure 21.11, making certain that both bars rest flat against the bottom of the vise and that they are butting against the edge of each vise jaw. Lift the carriage toward the key guide and cutter until a flat portion of the service bar rests against the key guide. (To lift the KD50A carriage, pull down and press the carriage release button between the vise jaws.)

Turn the cutter by hand. The machine is correctly adjusted if the cutter barely grazes the top of the right service bar. If the cutter is stopped from turning or turns freely without contacting the service bar, the cutting depth must be adjusted, as follows:

Loosen the Allen screw that holds the key guide.

Turn the cutting depth micrometer adjusting knob behind the guide, either left or right. This moves the key guide in or out. Do this until the cutter just grazes the top of the right service bar when the left service bar is resting against the key guide. Turn the cutter by hand; adjust to the high spot of the cutter.

Tighten the key guide Allen screw.

Note: This adjustment must be made if the cutter is replaced or whenever a test fails to work, indicating that the cutter may have worn down somewhat and resulting in cuts that are too shallow.

Adjusting for proper lateral distance (spacing)

Key cutting accuracy also depends on spacing the key and blank key the same as the distance between the key guide and cutter. To assure that the lateral distance adjustment is correct, refer to Figures 21.11 to 21.14 and proceed as follows.

Insert the service bars into the vise jaws, making sure that each service bar is butting against the edge of each vise jaw. *This is critical.*

Rotate the key setting gauge up, and make certain that both setting gauge shoulders rest *exactly* against the service bar stops as shown in Figure 21.12. If there is a discrepancy, loosen the right setting gauge Allen screw and adjust so that both gauge shoulders rest exactly against both service bar stops.

Figure 21.11 Clamp the two service bars into the vise jaws. (*Ilco Unican Corp.*)

Figure 21.12 Make certain both setting gauge shoulders rest against the service bar stops. (*Ilco Unican Corp.*)

Lift the carriage to the key guide and cutter. Insert the key guide and cutter into the V-shaped grooves in the service bars as shown in Figure 21.13. Both the key guide and the tip of a cutter wheel tooth must fit exactly into their V grooves or the setting will not be accurate (make certain that you do not seat the space between two cutter teeth into the V groove).

If the guide and cutter do not seat exactly into the V grooves, the distance between the cutter and guide must be altered. Loosen the Allen screw in the key guide assembly, and turn the micrometer adjusting knob fore or aft. This action shifts the position of the key guide assembly to the left or right. Continue until the key guide and the cutter both drop into the V notches of the service bars.

Insert the pattern key, left to right, into the left vise. Rotate the key setting gauge up and set its left shoulder against the shoulder of the pattern key. Be sure the key is lying flat along the bottom of the vise. Secure the key by turning the clamp assembly clockwise.

Insert the key blank in the same manner into the right vise and secure. Make sure that the key setting gauge is exactly against both key shoulders. The key and key blank now are spaced the correct distance apart, and they are ready for cutting (Figure 21.14).

Figure 21.13 Insert key guide and cutter into the V-shaped grooves in the service bars. (*Ilco Unican Corp.*)

Figure 21.14 The key and key blank are now spaced the correct distance apart for cutting. (*Ilco Unican Corp.*)

Aligning keys without shoulder (Ford and Best)

On keys without shoulders, the key setting gauge cannot be used. It is necessary to use the service bar to correctly position the key and the blank.

The vise jaws have a series of slots that can be used for the service bar. Also, note the key head rest (KD50A-60), which prevents the key from tilting as the vise jaw tightens. The key head rest can be moved to properly support the key (Figure 21.15).

Aligning narrow blade cylinder keys

Some keys have a very narrow blade; therefore they sit deep in the vise jaws with only part of the cuts showing above the vise. This makes it necessary to use the service pins to raise the key for proper cutting.

Figure 21.15 The key head rest can be moved to properly support the key. (*Ilco Unican Corp.*)

Insert an equal-size pin under each key and blank on the bottom of the vise jaws. This raises both the key and blank to make the correct depth of cut (Figure 21.16). Do not cut into the vise jaw.

Aligning double-sided cylinder keys

Before cutting this style of key, examine the key to see if there is a milled groove on either side. If so, then reverse the vise jaw and clamp the key using the V jaws. The key will be held securely when only the top or bottom V jaw fits into a milled groove. When there is no V groove on either side of the key, then use the flat vise jaw.

If the cuts are not the same on both sides of the key, make the shallow cuts first. In this way, when you turn the key over to cut the second side, there will be enough metal to grip the key securely during the actual cutting. To reverse the vise jaw, loosen the retaining screws at the base of the vise jaws. Raise, rotate, and reseat both vise jaws; then tighten their retaining screws. Note the V shape of the jaws. Insert the key between the jaws with a milling groove resting in the point of the V. This holds the blank securely. Align for spacing and proceed out.

Aligning carriage to prevent vise jaw damage

This machine is equipped with a carriage stop that prevents the carriage from moving all the way up to the cutter. When properly adjusted, it stops the cutter from grinding into the vise jaw. Such a condition could occur when reaching the tip of the cut key and the carriage lever continues to move the carriage.

Equal size service pins must rest flat along bottom of each vise with keys resting flat on top of pins so that keys are raised toward key guide and cutter.

Figure 21.16 Insert pin under key and key blank to allow correct depth of cut to be made. (*Ilco Unican Corp.*)

Figure 21.17 The carriage stop is a U-shaped channel secured to the housing by set screws. (*Ilco Unican Corp.*)

The carriage stop (Part No.KD50A-144) is a U-shaped channel secured to the housing by set screws. It is possible to span the travel of the carriage during the cutting cycle; normally, this position does not change. In addition, there is a carriage stop adjusting screw that is installed in the carriage; this screw controls the distance between the cutter and vise jaw (Figure 21.17).

The carriage stop adjusting screw is set at the factory to create a clearance of −0.005 inch between the cutter and the vise jaw (Figure 21.18). This distance is not critical and can be set without measuring instruments. Just loosen the lock nut and turn the screw in or out so the cutter does not touch the vise jaw. The machine should be off. When an ordinary business card can slide between the cutter and vise jaw, the adjustment is correct and the accuracy of key cutting will not be affected. *Caution:* Do not make this clearance too wide; key cutting could be affected on some keys having deep cuts.

It's a good idea to check the clearance on a regular basis, especially when a large quantity of keys are cut. If the cutter is allowed to strike the vise jaw, the edges of the cutter will be dulled immediately, causing a reduction in the life of the cutter.

Some important tips

- Clean vise jaws regularly so that no metal chips lie under the keys. It is essential that both keys lie flat across the entire width of both vise jaws. Neither key should be tilted. Do not use pliers or other tools to tighten the vise jaws. Firm hand pressure is sufficient.

- Keep the carriage shaft free of metal chips. A thin film of oil can be applied to it. The carriage should travel smoothly along its shaft.

- Never touch the shoulder of a key to the side of the key guide; this will cause the shoulder of a key blank to touch the side of the cutting wheel. When this happens, some of the metal will be cut away from the shoulder of the key guide. If the resulting duplicated key is duplicated two, three, or four times over, an error will accumulate and cause a nonoperating key. Do not grind away the shoulder.

- Don't run the cutter into the vise jaw; this will only dull the cutter and reduce cutter efficiency.
- Keep the cutter clean. Don't let any foreign objects or instruments blunt it. This cutter is a precise cutting tool. Handle it with care.
- Lubrication of moving parts is important. Oil cups are provided to keep the cutter shaft bearings well lubricated. The carriage spindle should be lubricated with a thin film of oil.

Figure 21.18 The carriage stop adjusting screw is set at the factory. (*Ilco Unican Corp.*)

22

Test Your Knowledge

This test is based on the general knowledge section of the International Association of Home Safety and Security Professionals' Registered Professional Locksmith certification program. If you can earn a passing score, you should be able to pass the RPL test and many other locksmithing certification and licensing examinations.

When answering the questions, choose the one best answer for each question. If more than one answer is chosen for any question, you will receive no credit for answering that question.

1. A Schlage Primus cylinder has:
    a. 12 lever tumblers
    b. 16 rotating disc tumblers
    c. finger pins
    d. rotating ward rings
2. When drilling a cylinder open:
    a. position the drill bit at least ½ inch above the cylinder and drill upward.
    b. position the drill bit at the shear line.
    c. position the drill bit at least ½ inch below the cylinder and drill downward.
    d. position the drill at the first letter on the face of the cylinder.
3. Chicago Lock Company
    a. made the first tubular key lock.
    b. made the first lever tumbler lock.
    c. makes the Chicago "Stamper-bolt" lock.
    d. made the first key-in-knob lock.
4. When viewed from the exterior side, a door that opens inward and has hinges on the right:
    a. a right-hand door
    b. a left-hand door

    c. a left-hand reverse bevel door

    d. a right-hand reverse bevel door

5. When shimming a cylinder open:

    a. insert the shim into the keyway.

    b. insert the shim along the bottom of the keyway.

    c. insert the shim along the left side of the plug.

    d. insert the shim between the plug and cylinder housing at the pin tumblers.

6. To earn the UL burglary safe rating TRTL-30, a safe:

    a. must weigh at least 1000 pounds.

    b. must have a door that opens inward.

    c. must have relockers along its floor.

    d. must weigh at least 750 pounds, or be equipped with anchors.

7. A utility patent:

    a. relates to a product's appearance, is granted for 14 years, and is renewable.

    b. relates to a product's function, is granted for 17 years, and is nonrenewable.

    c. relates to a product's appearance, is granted for 17 years, and is renewable.

    d. relates to a product's function, is granted for 35 years, and is nonrenewable.

8. To earn a UL-437 rating, a lock must:

    a. pass an attack test using common hand and electric tools, drills, saw blades, puller mechanisms, and picking tools.

    b. pass a performance test only.

    c. use a patented key.

    d. use hardened-steel mounting screws.

9. The pin tumbler lock patented by Linus Yale Jr. in 1865 used a mortise style cylinder that operated with a flat key.

    a. true

    b. false

10. In 1868 Henry R. Towne:

    a. joined with Linus Yale Jr. to form the Yale Lock Manufacturing Company.

    b. joined with Linus Yale Sr. to form the Yale and Towne Manufacturing Company.

    c. joined with Walter Schlage to form the Schlage Lock Company.

    d. joined with Stephen Bucknall to form the Eagle Lock Company.

11. A money safe is designed to protect currency from burning during a fire.

    a. true

    b. false

12. A "lock" is:

    a. any barrier or closure.

    b. any fastening device that allows a person to open and close a door, window, cabinet, drawer, or gate.

    c. any device that incorporates a bolt, cam, shackle, or switch to secure an object—such as a door, drawer, or machine—to a closed, locked, on, or off position, and that provides a restricted means of releasing the object from that position.

    d. any device or object that restricts entry to a given premise.

13. Tumblers are:
    a. fixed projections on the case of a lock.
    b. small objects that protrude out of a lock's cam to activate the bolt.
    c. small pins, usually made of metal, that move within a bit key lock's case to prevent unauthorized keys from entering the keyhole.
    d. small objects, usually made of metal, that move within a lock cylinder in ways that obstruct a lock's operation until an authorized key or combination moves them into alignment.
14. Parts within the pin chambers of a typical pin tumbler cylinder include top springs, top pins, bottom pins, and bottom springs.
    a. true
    b. false
15. Electric switch locks:
    a. are mechanical locks that have been modified to operate with ac electricity instead of with a key.
    b. are mechanical locks that have been modified to operate with battery power.
    c. are installed in metal doors to give electric shocks to intruders.
    d. complete and break an electric current when an authorized key is inserted and turned.
16. Grade 3 locks (based on ANSI A156 standards) are for heavy-duty commercial applications.
    a. true
    b. false
17. Worn pin tumblers are generally easier to pick than are new pin tumblers.
    a. true
    b. false
18. Disc tumbler (wafer) locks have two shear lines that tumblers can protrude past to prevent a plug from being rotated.
    a. true
    b. false
19. When picking a typical pin tumbler cylinder:
    a. the bottom pins must be raised to the shear line.
    b. the top pins must be pulled below the shear line.
    c. the tension wrench (turning tool) should be inserted only after all the pins have been picked into a position that frees the plug to turn.
    d. the pick should be inserted into the keyway only after the tension wrench (turning tool) has rotated the plug into position and has been removed from the keyway.
20. Before duplicating a General Motors VATS key:
    a. you have to remove the resistor pellet from the key being copied.
    b. you have to demagnetize the resistor pellet in the key being copied.
    c. you need to determine the resistance of the pellet in the key to determine which blank to use.
    d. use a ground strap to avoid transferring static electricity to the resistor pellets.

21. When picking a pin tumbler cylinder, sometimes spraying a little lubrication into the keyway can be helpful.
    a. true
    b. false

22. Before impressioning a pin tumbler cylinder, it's usually helpful to thoroughly lubricate the pin chambers.
    a. true
    b. false

23. The older a pin tumbler is, the easier it will be to impression.
    a. true
    b. false

24. The most popular type of lock used on GM cars is a:
    a. side bar lock
    b. pin tumbler lock
    c. tubular key lockd.
    d. flat key lock

25. If a cut in a key for a pin tumbler cylinder isn't deep enough:
    a. the bottom pin for that cut will drop below the shear line
    b. the top pin for that cut will drop below the shear line
    c. the bottom pin for the adjacent cut will drop below the shear line
    d. the bottom pin for that cut will rise above the shear line

26. The bit of a bit key:
    a. manipulates the tumblers in a warded bit key lock
    b. enters the keyway of a cylinder lock
    c. enters the keyhole of a bit-key lock
    d. comes into contact with the tailpiece of a bit key lock to extend and retract the lock bolt

27. Spool and mushroom pins:
    a. make a lock easier to pick
    b. make keys harder to duplicate
    c. make a lock harder to pick
    d. make keys easier to duplicate

28. In a typical dial combination lock, each tumbler has either three or four gates.
    a. true
    b. false

29. Usually the simplest way to change the combination of a double-bitted cam lock is to:
    a. rearrange the positions of two or more tumblers
    b. remove two tumblers and replace them with new tumblers
    c. remove the tumbler assembly and replace it with a new one
    d. connect a new tumbler assembly to the existing one

30. The blade length of a key is most indicative of the:
    a. manufacturer of the lock the key is designed to operate
    b. number of tumblers in the cylinder the key is designed to operate
    c. lengths of the master pins in the cylinder the key is designed to operate
    d. number of master pins in the cylinder the key is designed to operate

31. An uncut key is called a:
    a. key blank
    b. change key
    c. master key
    d. bit key
32. The "hand of a door" refers to the:
    a. location of the hinges and swing of the door
    b. type of lock on the door
    c. doorknob side of a door
    d. direction that the doorknob turns to open the door
33. Which type of lock generally provides the least security?
    a. pin tumbler
    b. deadbolt
    c. side bar
    d. warded
34. Which type of lock would most likely provide the most security?
    a. ANSI Grade 3 key-in-knob lock with dual pin tumbler cylinders, five master pins, and hardened steel bolt
    b. ANSI Grade 2 key-in-knob with dual UL-437 listed cylinders, five master pins, and hardened steel bolt
    c. ANSI Grade 2 deadbolt lock with UL-437 listed cylinder, patented key, and hardened steel bolt
    d. Interlocking deadbolt with spool and mushroom bottom pins, five master pins, and a key stamped "Do Not Duplicate"
35. Among common keys, those with the longest life span are made of:
    a. aluminum
    b. brass
    c. copper-silver alloy
    d. borax iron-steel alloy compounds
36. Which of the following key combinations provide the least security?
    a. 36242
    b. 24263
    c. 11111
    d. 22241
37. Which of the following key combinations provide the most security?
    a. 352146
    b. 24263
    c. 36242
    d. 11111
38. When cutting a lever tumbler key by hand, the first cut should be the:
    a. throat cut
    b. tip cut
    c. stop cut
    d. lever cut
39. Successful lock picking depends primarily on:
    a. length of tension wrench

b. material of cylinder

c. length of lock bolt

d. knowledge about lock

40. James Sargent is famous for:

a. being the first person to pick a Yale pin tumbler cylinder

b. inventing the double-acting lever tumbler lock

c. inventing the time lock for banks

d. being the first person to pick a double-acting lever tumbler lock

41. Which manufacturer is best known for its pushbutton combination locks?

a. Kwikset Corporation

b. Best Lock Corporation

c. Better Lock and Interchangeable Core Company

d. Simplex Access Controls

42. Which manufacturer is best known for its interchangeable core locks?

a. Kwikset Corporation

b. Best Lock Corporation

c. Better Lock and Interchangeable Core Company

d. Simplex Access Controls

43. Which manufacturer is best known for its residential key-in-knob and deadbolt locks?

a. Kwikset Corporation

a. Best Lock Corporation

c. Better Lock and Interchangeable Core Company

d. Simplex Access Controls

44. As a safety precaution, before servicing an ignition lock on an automobile a locksmith should:

a. disconnect the negative battery cable from the battery

b. insert the proper key into the ignition lock

c. connect a ground cable to the automobile's battery

d. touch a plastic part of the vehicle to discharge static electricity

45. Ford began using wafer side bar locks in their automobiles in which year?

a. 1984

b. 1975

c. 1981

d. 1990

46. Which code series is commonly used by Ford glove compartment and trunk locks?

a. "FMC" 2223–9999

b. "FB" 0001–1863

c. "GT" 1111–9999

d. "FGT" 001–1888

47. General Motors codes follow the following rule:

a. no two adjacent cuts have a depth difference of more than one

b. if all six depths of a key are added together, they equal an odd number

c. no two adjacent cuts have a depth difference of more than two

d. all codes consist of either all even numbers or all odd numbers

48. General Motors ignition lock codes can generally be found:
    a. on the ignition lock
    b. on the passenger side door lock
    c. below the Vehicle Identification Number (VIN) on the vehicle's engine
    d. under the vehicle's break pedal
49. How many styles of lock pawls does General Motors use in its various car lines?
    a. three
    b. more than twenty
    c. two
    d. five
50. Which code series is commonly used on Chrysler door and ignition locks?
    a. EP 1–3000
    b. CHR 1–5000
    c. CRY 1–4000
    d. GM 001–6000
51. Prior to 1995, 15 different resistor values were used for VATS or PASSKey systems.
    a. true
    b. false
52. "Nonvolatile" memory refers to a type of memory that isn't automatically erased from the memory of an electronic system during a power failure.
    a. true
    b. false
53. The Schlage J key blank enters the E, EF, F, and P keyways.
    a. true
    b. false
54. The Abus Diskus padlock has an arced shackle.
    a. true
    b. false
55. A safe's spline key holds the drive cam to the spindle.
    a. true
    b. false
56. Schlage Lock Company uses a key cut depth increment of 0.0002 inch.
    a. true
    b. false
57. The Schlage D Series lockset is designed primarily for light residential applications.
    a. true
    b. false
58. Southern Steel is a major manufacturer of prison/detention locks and hardware.
    a. true
    b. false
59. The rotating constant method of masterkeying requires change keys to have at least four bittings in the same position shared with the TMK.
    a. true
    b. false
60. A typical disc tumbler lock has at least eight different tumble depths.
    a. true
    b. false

Test Answers

When scoring the test, give yourself two points for each correctly answered multiple choice question and one point for each correctly answered true/false question. In each case where you gave no answer to a question and in each case where you gave two or more answers to a question, give yourself no credit.

Scoring Key: 75% is minimum for passing; 85% is very good; 95% or better is excellent.

Answers

1. c	21. a	41. d
2. b	22. b	42. b
3. a	23. b	43. a
4. a	24. a	44. a
5. d	25. d	45. a
6. d	26. c	46. b
7. b	27. c	47. c
8. a	28. b	48. a
9. a	29. c	49. b
10. a	30. b	50. a
11. b	31. a	51. a
12. c	32. a	52. a
13. d	33. d	53. a
14. b	34. c	54. a
15. d	35. b	55. a
16. b	36. c	56. b
17. a	37. a	57. b
18. a	38. a	58. a
19. a	39. d	59. b
20. c	40. c	60. b

Lock and Safe Manufacturers

A-1 Security Manufacturing Corporation
3528 Mayland Ct.
Richmond, VA 23233
(804) 747-0095

Makes more than 100 locksmithing tools. Brand names: PAK-A-PUNCH, TB3, THE BLOCK.

Abloy Security, Inc.
6015 Commerce, Suite 450
Irving, TX 75063
(800) 367-4598

Makes high-security locks.

Allied Gary International
P. O. Box 947 Gary Way
Waynesboro, GA 30830
(800) 456-4279; (706) 544-1918;
Fax: (706) 554-3319

Makes "made to order" money chests, UL-listed safes, bank equipment, and other security products.

Almont Lock Co., Inc.
113 School St.
Almont, MI 48003
(810) 798-8950

Makes rekeyable brass padlocks and cylinders.

American Lock Company
3400 W. Exchange Rd.
Crete, IL 60417
(708) 534-2000; Fax: (708) 534-0531

Makes solid brass and solid steel padlocks, hasps, cam locks, and other lock products. Background: Founded in 1912 by John Junkunc as "Junkunc Safe and Lock Company."

American Security Products Company (or AMSEC)
11925 Pacific Ave.
Fontana, CA 92335
(909) 685-9680

Makes a wide variety of safes under the brand names AMSEC, Star, and Major. Background: Founded in 1946 as Star Safe Company (or Starco). In 1964, the company changed its name to American Security Products Company. In 1987, the company purchased one of its biggest competitors, Major Safe Company.

ASSA High Security Locks
10300 Foster Ave.
Brooklyn, NY 11236
(718) 927-2772; Fax: (718) 257-2772

A leading manufacturer of high-security locks. Trade names: ASSA, ASSA Twin 6000, Insta-Core. Background: A subsidiary of Securitas Lock Group Incorporated.

Baton Lock & Hardware Company Inc.
14275 Commerce Dr.
Garden Grove, CA 92643
(714) 265-3636; Fax: (714) 265-3630

Makes padlocks, cam locks, and assorted door hardware.

Black & Decker
516 E. Santa Ana St.
Anaheim, CA 92803
(714) 999-2362; Fax: (714) 999-2290

A leading manufacturer of locks, door hardware, and tools. Background: Purchased Emhart Corporation in 1989.

CCL Security Products
199 Whiting St.
New Britain, CT 06051
(203) 229-6199; Fax: (203) 223-7601

Makes combination padlocks, cam locks, and other types of locks and security products. Brand names: Sesamee and Prestolock. Background: Corbin Cabinet Lock (CCL) was for many years a division of Emhart Industries. In 1987 the Eastern Company bought the division and reorganized it under the name "CCL Security Products."

Corbin Russwin Architectural Hardware
225 Episcopal Rd.
Berlin, CT 06037
(203) 225-7411; Fax: (203) 828-7266

Makes a complete line of door hardware products, including locks, exit devices, door closers, and key systems. Background: In May 1993 Corbin and Russwin brands merged.

Chicago Lock Company
4311 W. Belmont
Chicago, IL 60641
(312) 282-7177; Fax: (312) 282-8199

Makes tubular-key locks and assorted locks and keys. Trade names: Ace and AceII.

Diebold, Inc.
P. O. Box 8230
Canton, OH 44711-8230
(216) 798-5625; Fax: (216) 798-5626

Makes safes. Trade name: Banklock.

Dom Security Locks N.A.
100 Central Ave.
Brockville, Ontario
Canada K6V-5W6
(800) 363-4803; Fax: (613) 498-6820

DLA Security Systems Inc.
629 Kimball Ave.
Westfield, NJ 07090
(908) 233-7755

Makes masterkeying software.

Federal Lock Company
1320 Enterprise Dr.
Romeo, IL 60441
(708) 378-4745; Fax: (708) 378-4767

Makes padlocks and high-security hasps. Background: Founded in 1992 by Ken Erickson and some of his associates.

Folger Adam Company
16300 West 103rd St.
Lemont, IL 60439
(708) 739-3900; Fax: (708) 739-6138

A leading manufacturer of electric strikes, electric locks, detention hardware, and assorted security products.

Framon Manufacturing Company, Inc.
909 Washington Ave.
Alpena, MI 49707
(517) 354-5623; Fax: (517) 354-4238

Makes key cutting and code machines.

Gardall Safe Corporation
219 Lamson St.
Syracuse, NY 13206
(800) 722-7233; (315) 432-9115;
Fax: (315) 434-9422

Makes a wide range of UL-listed safes for commercial and residential use.

Ilco Unican Corporation
400 Jeffreys Rd.
Rocky Mount, NC 27804
(919) 446-3321; Fax: (919) 446-4702

A leading manufacturer of a wide range of locks and security products.

InstaKey Lock Corp.
3021 S. Tejon St.
Englewood, CO 80110
(303) 761-9999

Makes high-security locks.

International Locking Devices, Ltd.
104-4 Branford Rd.
North Branford, CT 06471
(800) 863-9600; (203) 481-6738;
Fax: (203) 481-8490

Makes assorted gate locks and latches.

Jet Hardware Manufacturing Corporation
800 Hinsdale St.
Brooklyn, NY 11207
(718) 257-9600; Fax: (718) 257-0973

Keedex Manufacturing
12931 Shackelford Lane
Garden Grove, CA 92641-5108
(714) 636-5657; Fax: (714) 636-5680

Makes locksmithing tools and equipment.

Kustom Key Inc.
1010 Aviation Dr.
Lake Havasu City, AZ 86403
(800) 537-5397; Fax: (800) 235-4728

Makes neuter bow key blanks incised or embossed with company names or custom messages.

Kwikset Corporation
One Park Plaza, Ste. 1000
Irvine, CA 92714
(714) 474-8800; Fax: (714) 474-8879

A leading manufacturer of deadbolt locks, key-in-knobs, lever locks, handlesets, and other security hardware. Major brand and trade names: Kwikset and Titan. Over 2000 employees. Operating facilities in Anaheim, CA; Bristow,

OK; Denison, TX; and Irvine, CA. Background: Founded in the late 1940s by Adolf Schoepe and Karl Rhinehart as a joint partnership and named "Gateway Manufacturing Company." In 1947 Gateway Manufacturing changed its name to Kwikset. In 1952 Kwikset became publicly owned; about 50 percent of its stock was placed on the market. In July 1957 the American Hardware Corporation acquired controlling interest in Kwikset. By 1958 Kwikset became an operating division of American Hardware. In 1964 the American Hardware Corporation merged with Emhart Manufacturing Company, and established the Emhart Corporation. In 1989 the Black & Decker Company purchased the Emhart Corporation, including the Kwikset division.

Locksoft
P. O. Box 3168
Bristol, TN 37625-3168
(800) LOCKSOFT; (615) 652-2454;
Fax: (615) 652-2797

Makes computer software for locksmiths.

M.A.G. Eng. & Mfg. Inc.
15261 Transistor Lane
Huntington Beach, CA 92649-1141
(800) 624-9942; (714) 891-5100;
Fax: (714) 892-5845

A leading manufacturer of high-security strike boxes, door reinforcers, and other door and window hardware.

Major Manufacturing, Inc.
P. O. Box 788
Artwood, CA 92601
(714) 772-5202; Fax: (714) 772-2302

Makes locksmithing tools and equipment.

Master Lock Company
2600 North 32nd St.
Milwaukee, WI 53210.
(414) 444-2800; Fax: (414) 444-0322

A leading manufacturer of padlocks, key-in-knobs, deadbolts and other types of locks and related products. Major brand and trade names: Durashine finish and The Pro Series line of weather-resistant padlocks. Over 1800 employees at its 800,000 square-foot manufacturing plant (warehouse) headquarters in Milwaukee and at its production facility in Auburn, AL. Background: Founded in 1921 by Harry E. Soref, who introduced the Master Laminated Keyed Padlock. In 1970, Master Lock became a subsidiary of American Brands Inc. In 1985, Master Lock acquired Dexter Lock Company. In 1988, American Brands formed Master-Brand Industries, Inc., a subsidiary for its hardware and home improvement products firm—including Master Lock. In 1990, Master Lock International was formed. Master Lock started its Door Hardware Division and introduced a line of easy-to-install door hardware in 1991. In the same year, Master Lock International opened an office in Paris, France, and a warehouse in Rotterdam, the Netherlands.

MAXIS Security Locks
103 Powell St.
Vancouver, BC V6A-1H9
(604) 255-2005

Medeco Security Locks, Inc.
P. O. Box 3075
Salem, VA 24153
(703) 380-5000; Fax: (703) 380-5010

A leading manufacturer of high-security locks and cylinders.

Monarch Tool & Mfg. Co., Inc.
P. O. Box 427
Covington, KY 41012
(800) 462-9460; (606) 261-4421;
Fax: (606) 261-7403

Makes coin boxes and assorted vending machine products. Background: Established in 1903.

Mosler
1561 Grand Blvd.
Hamilton, OH 45012
(513) 867-4000; Fax: (513) 867-4062

A leading manufacturer of UL-listed burglary safes, vaults, vault doors, safe deposit boxes, and related products. Background: Established in 1867.

Mul-T-Lock USA Inc.
300-1 Route, 17S, Suite A
Lodi, NJ 07644
(201) 788-3222

Makes high-security locks.

The New England Lock & Hardware
  Company, Inc.
P. O. Box 544
South Norwalk, CT 06856
(203) 866-9283; Fax: (203) 838-4837

Makes assorted vertical deadbolts, rim locks, and other locks and cylinders. Brand name: SEGAL.

Ohio Travel Bag Mfg. Co.
811 Prospect Ave.
Cleveland, OH 44115
(216) 621-5963; (216) 621-5559

Makes luggage and trunk keys, locks, and hardware.

Preso-Matic Keyless Locks
237 Coastline Rd.
Sanford, FL 32771-6659
(407) 324-9933; Fax: (407) 328-9977

Makes mechanical pushbutton door locks for commercial and residential installations.

Qualtec Data Products, Inc.
47767 Warm Springs Blvd.
Fremont, CA 94539
(800) 628-4413; (510) 490-8911;
Fax: (510) 490-8471

Makes assorted locks and anti-theft devices for personal computers. Background: established in 1984.

Rofu International Corporation
3725 Old Conejo Rd.

Newbury Park, CA 91320
(800) 255-ROFU; (805) 499-0316;
Fax: (805) 499-3154

A leading manufacturer of electric strikes, electromagnetic locks, and other security products.

Sargent & Greenleaf, Inc.
One Security Dr.
Nicholasville, KY 40356
(606) 885-9411; Fax: (800) 634-4843

A leading manufacturer of locks for safes, vaults, and safe deposit boxes. Occupies 10,000 square feet of manufacturing space and 22,000 square feet of offices. The company also serves European and other international customers from a facility in Switzerland. Background: In 1857, James Sargent, then an employee of Yale & Greenleaf Lock Manufacturers, devised a magnetic bank lock that quickly gained acceptance from safe manufacturers and the United States Treasury Department. That lock brought financial stability to the company and paved the way for James Sargent to continue devising new security products. In 1865, one of Mr. Sargent's former employees, Halbert Greenleaf, became his equal partner in the firm of Sargent & Greenleaf.

Schlage Lock Co.
2401 Bayshore Blvd.
San Francisco, CA 94134
(415) 467-1100; Fax: (415) 330-5626

Makes assorted door locks and security products.

Securitech Group Inc.
54-45 44th St.
Maspeth, NY 11378
(800) 622-LOCK; (718) 392-9000; Fax: (718) 392-8944

Makes high-security door locks and related products. Brand names include: MAGLATCH, MP Police Lock, Double-throw, and 1derbolt.

Security Control Systems
20202 S. Holly
Frankfort, IL 60423
(815) 469-0532; Fax: (815) 469-6425

Makes GSA-approved key control software.

Select Hardware Corporation
675 Emmett St.
Bristol, CT 06010
(203) 589-7303; Fax: (203) 589-6245

Makes deadbolts that can be fitted with many popular key-in-knob or interchangeable core cylinders.

Sentry Group
900 Linden Ave.
Rochester, NY 14625
(800) 828-1438; Fax: (716) 381-8559

A leading manufacturer of UL-listed fire safes and containers for the home and office.

Silca Keys U.S.A., Inc.
9049 Dutton Dr.
Twinsburg, OH 44087
(216) 487-5454; Fax: (216) 487-5459

Makes keys and key machines.

Simplex Access Controls
2941 Indiana Ave.
P. O. Box 4114
Winston-Salem, NC 27115-4114
(910) 725-1331; Fax: (800) 346-9640

Southern Steel
4634 S. Presa St.
San Antonio, TX 78223
(210) 533-1231; Fax: (210) 533-2211

A leading manufacturer of detention locks, hinges, and other door hardware.

Trine Products Company
1430 Ferris Pl.
Bronx, NY 10461
(718) 829-4796; Fax: (718) 792-9127

Makes electric strikes and other products.

Ultrak, Inc.
2400 Industrial Lane, Suite 2000A
Broomfield, CO 80020
(800) 846-8884; (303) 466-7333;
Fax: (303) 469-8336

A leader manufacturer of CCTV systems and equipment.

U.S. Lock Corporation
77 Rodeo Dr.
Edgewood, NY 11717
(800) 525-5000

Manufacturer and distributor of locks and other architectural hardware and security products.

Vaughan & Bushnell Mfg. Co.
11414 Maple Ave.
Hebron, IL 60034
(815) 648-2446; Fax: (815) 648-4300

A leading manufacturer of hammers, hatchets, axes, and assorted striking tools. Major brand and trade names: Doubleduty, Shock-Blok, Steel Eagle, and Supersteel. Background: Founded in 1869 in Peoria, Illinois by a young blacksmith named Alexander Vaughan. His only son, Sanford S. Vaughan, joined the company in 1886 and took over management in 1905. Sanford served the company for 65 years until his death in 1951. Members of the Bushnell family also held management positions with the company until their interest was purchased by the Vaughan family in 1922, the same year that Sanford's son, Howard Sr. joined the company. Howard A. Vaughan, Sr. was president of the company from 1945 to 1966. Since 1966, the company has been headed by Howard A. Vaughan, Jr., which makes the company one of the oldest family-run businesses in the United States.

Von Duprin, Inc.
2720 Tobey Dr.
Indianapolis, IN 46219
317) 897-9944; Fax: (317) 899-9302

A leading manufacturer of door exit devices, electric strikes, and related security products.

Weiser Lock
6660 So. Broadmoor Rd.
Tucson, AZ 85746

(800) 688-5625; (602) 741-6200;
Fax: (602) 741-6241

Makes assorted residential locks. Background: established in 1904.

Weslock National Inc.
13344 So. Main St.
Los Angeles, CA 90061
(310) 327-2770; Fax: (310) 324-4624.

Makes assorted residential locks.

Wilson Bohannan Co.
621 Buckeye St.
Marion, OH 43302
(614) 382-3639; Fax: (614) 383-1653

Makes solid brass, pin-tumbler padlocks. Employs nearly 70 people in a 50,000 square foot facility. Major brand and trade names: WB, and Top Brass. Background: Founded in 1860 by Wilson Bohannan as a machine shop; he ran the business out of the back of his home in Brooklyn, New York. In the 1880s, the business grew to the point where it had to move to its own four-story, 10,000 square foot building. In 1927, Wilson Tway, who was related to the Bohannan family through marriage, moved the company to Marion, Ohio. He remained president of the company until he died in 1982. His son, Dick Tway, was president until his untimely death in 1989. Dorthea Tway Norris and other family members maintain an active interest in the company.

X-10 (USA) Inc.
91 Ruckman Rd.
Closter, NJ 07624
(201) 784-9700; Fax: (201) 784-9464

A leading manufacturer of do-it-yourself home automation systems, security systems, and related devices. Major brand and trade names: X-10.

Yale Security Inc.
P. O. Box 25288
Charlotte, NC 28229-8010
(704) 283-2101

Makes locks, door closers, exit devices, and assorted door hardware.

Locksmith Suppliers
Profiles and Addresses

Accredited Lock Supply Co.
1161 Paterson Plank Rd.
Secaucus, NJ. 07094
(800) 652-2835
Fax: (201) 865-5031

Complete line of locks and supplies; some access control and other electronic security supplies; 25 employees; 35,000 square feet warehouse. Background: Founded in 1974 as a locksmith shop by Rudy Weaver, and his son, Ron.

Acme Wholesale Distributors, Inc.
P. O. Drawer 13748
New Orleans, LA 70185
(800) 788-2263
Fax: (504) 837-7321

Full-line distributor of most major lock lines, supplies, safes, and access control devices. Has branch distribution centers in New Orleans, Houston, San Antonio, and Ft. Worth. Background: Founded in the early 1970s. Joined the LSDA Group in 1985.

American Lock & Supply, Inc.
4411 E. LaPalma Ave.
Anaheim, CA 92807
(800) 844-8545; (714) 996-0791;
Fax: (714) 579-3508

Full-line distributor of most major lock lines. More than $20 million inventory at 11 locations throughout the United States.

Aristo Sales Company, Inc.
27-24 Jackson Ave.
Long Island City, NY 11101
(800) 221-1322; (718) 361-1040;
Fax: (718) 937-5794

Supplier of automotive locks, alarms, safes, supplies.

Arius
8259 Exchange Dr.
Orlando, FL 32809-7687
(407) 850-2625;
Fax: (407) 850-2616

Wide range of electronic security products, including CCTVs, access control, and burglar/fire alarms.

Armstrong's Lock & Supply, Inc.
1440 Dutch Valley Place, NE
Atlanta, GA 30324
(800) 726-3332; (404) 875-0136;
Fax: (404) 888-0834

Full-line distributor of locks, safes, supplies, and electronic security devices. Carries more than 13,000 different products from more than 250 manufacturers. Branches in Miami, Tampa, Jacksonville, and Norfolk.

Clark Security Products
7140 Engineer Rd.
San Diego, CA 92111-1491

More than 30,000 products from more than 250 manufacturers. Branches in San Diego, Sacramento, and Los Angeles, California; Denver, Colorado; Phoenix, Arizona; Seattle, Washington; Washington D.C.; and New England. Background: Founded in 1953.

Fried Brothers, Inc.
467 N. 7th St.
Philadelphia, PA 19123
(800) 523-2924
Fax: (215) 523-1255

Products from more than 120 different manufacturers. Background: Established in 1922.

KingAlarm
35 Green St.
Hackensack, NJ 07601
(800) 526-0162; (201) 488-4990
Fax:(201) 342-7097

Distributor of burglar/fire alarm systems, CCTV equipment, access control equipment, telephone systems, intercoms and related accessories and supplies. Stocks more than 9000 products. Warehouse sales centers located in: Elmsford, New York; Toms River, New Jersey; Kenilworth, New Jersey; Bellmawr, New Jersey; Anaheim, California; Doraville, Georgia; Farmington Hills, Michigan; and Deerfield Beach, Florida.

Maziuk & Company, Inc. 125
1 W. Genesee St.
Syracuse, NY 13204
(800) 777-5945; (315) 474-3959;
Fax: (315) 472-3111

Stocks wide range of locks and supplies from more than 135 manufacturers. Has a 5300 square foot branch office and warehouse in the Buffalo, New York area. Background: Established in 1943.

McDonald Dash Locksmith Supply, Inc.
5767 East Shelby Dr.
P. O. Box 752506-2506
Need City & State
(901) 797-8000
Fax: (901) 366-0005

Distributor of more than 15,000 locks, tools, supplies, and related security products from 125 manufacturers. Background: Established in 1945.

M. Taylor Inc.
5635-45 Tulip St.
Philadelphia, PA 19124
(800) 233-3355
Fax: (215) 288-2511.

More than 11,000 items from more than 130 manufacturers. Background: Established in 1912.

Security Lock Distributors
59 Wexford St.
P. O. Box 815
Needham Heights, MA 02194
(800) 847-5625
Fax: (800) 878-6400

Sells electronic and mechanical locking devices.

Wilco Supply
5960 Telegraph Ave.
Oakland, CA 94609
(800) 745-5450
Fax: (510) 653-5397

More than 32,000 different locks, tools, supplies, and related products. 52 employees. Background: Founded in 1951.

Wm. B. Allen Supply Co., Inc.
Allen Square
301 North Rampart St.
New Orleans, LA 70112-3106
(800) 535-9593
Fax: (800) 444-1726

Sells burglar/fire alarm systems, equipment, and supplies. Stocks more than 72,000 different products from more than 300 manufacturers. Background: Established in 1940.

Zipf Lock Co.
830 Harmon Ave.
Columbus, OH 4322
(800) 848-1577; (614) 228-3507
Fax: (800) 228-6320

Stocks more than 16,000 different locks, lock parts, tools and supplies. Background: Established in 1908.

Miscellaneous Important Addresses/Phone Numbers

Builders Hardware Manufacturers Association, Inc.
355 Lexington Ave., 17th Floor
New York, NY 10017
(212) 370-9047

Trade Association. Write for current information about ANSI standards for locks and other door hardware.

Internal Revenue Service (IRS)
Tax Help Line (800) 829-1040

Call for information about taxes or for tax-related publications.

International Association of Home Safety and Security Professionals, Inc.
P. O. Box 2044
Erie, PA 16512-2044

Trade Association for locksmiths, alarm installers, home automation system installers, security officers, and others. Write for current statistical information related to the security industry.

Lock Museum of America
P. O. Box 104
Terryville, CT 06786-0104
(203) 589-6359

Contact for information related to lock and key history.

Public Access Section Civil Rights Division
US Department of Justice
P. O. Box 65860
Washington, DC 20035-5860

Write for a free packet of information about the Americans with Disabilities Act (ADA). Specify whether you want the information in Braille, large print, audio tape, or computer disk.

Small Business Administration (SBA)
1441 L St., NW
Washington, DC 20461
(800) 827-5722

Contact for information about starting or running a small business.

Underwriters Laboratories Inc.
333 Pfingsten Rd.
Northbrook, IL 60062

Write for current information on UL standards or to find out if a manufacturers' product has a UL listing.

Locks, lock parts, tools and supplies. Background: established in 1908.

Plug Follower and Holder Diameters for Popular Locks

Manufacturer	Diameter (in inches)			
	0.395	0.495	0.500	0.550
Acrolock		X		
Corbin		X		
Corbin oversize				X
Corbin small pin	X			
Eagle		X		
Eagle small pin	X			
ILCO		X		
ILCO 4019 rim cylinder			X	
ILCO "peanut" cylinder	X			
Keil			X	
Kwikset		X		
Lockwood		X		
National (Rockford)		X		
National (Ozone Park)		X		
Russwin oversize				X
Sargent		X		
Segal		X		
Taylor		X		
Weslock (except KNK)		X		
Yale		X		
Yale small pin	X			

Glossary

Many locksmithing terms have popular synonyms and variant spellings. The entries in this glossary are based on a survey of locksmithing trade journals, reference books, technical manuals, supply catalogs, and trade association glossaries. When two or more synonyms or spellings are commonly used, they're listed in order of popularity. When two or more are equally popular, they are listed in order of age (from oldest usage to newest). In cases where two or more common locksmithing terms are spelled alike but have different meanings, each term is listed in a separate entry.

absolute master key   The key that operates every masterkeyed lock in a large or complex master key system.

ac   Alternating current. For practical purposes, alternating current is the type of electricity that flows throughout a person's house and is accessed by the use of wall sockets.

Ace lock   Sometimes used to refer to any tubular key lock; but the term is a trade name for tubular locks made by Chicago Lock Company.

A.D.A.   abb. Americans With Disabilities Act. A congressional act passed to ensure the rights of persons with various types of disabilities. Part of the act requires that buildings provide easy access to the physically disabled.

AHC   abb. Architectural Hardware Consultant (as certified by DHI).

ALOA   abb. Associated Locksmiths of America.

ampere   (or amp) A unit of electrical current.

annunciator   A device, often used in an alarm system, that flashes lights, makes noises, or otherwise attracts attention.

ANSI   abb. American National Standards Institute.

anti-pick latch   A spring latch fitted with a parallel bar that is depressed by the strike when the door is closed. When the bar is depressed, it prevents the latch from responding to external pressure from tools such as a shove-knife or an ice pick.

armored front   A plate covering the bolts or setscrews holding a cylinder to its lock. These bolts are normally accessible when the door is ajar.

ASIS   abb. American Society for Industrial Security.

backplate   (rim cylinder) A small plate applied to the inside of a door through which the cylinder connecting screws and bar is passed.

backset   (of a lock) The horizontal distance from the edge of a door to the center of an installed lock cylinder, keyhole, or knob hub. On locks with rabbeted fronts, it's measured from the upper step at the center of the lockface.

barrel key   A key with a bit projecting from a hollow cylindrical shank. The hollow fits over a pin in a lock keyway and helps keep the key aligned. The key is also known as a hollow post key or pipe key.

bell crank   (or bellcrank) (within the walls of an automobile's door) A flat metal plate attached to connecting rods. The bell crank converts movement from vertical to horizontal or vice versa.

BHMA   abb. Builders Hardware Manufacturers Association.

bicentric cylinder   A lock cylinder having two plugs, with a set of tumblers in each plug. The correct key can be inserted into either plug to open the lock. This permits masterkeying without reducing the security of the locking mechanism.

bit   That part of a bit key that enters a keyhole to operate a lock.

bit key   (or bit-key or skeleton key) A key with a bit projecting from a solid cylindrical shank. A bit key is used to operate a bit key lock.

bit key lock   A lock operated by a bit key lock.

bitting   (or cuts) A cut, or series of cuts, on the bit or blade of a key.

blank   (or key blank) A key that has not been cut or shaped to operate a specific locking mechanism.

bored lock   Any lock that is installed by cross-boring two holes in a door; one hole in the face of the door, and the other through the edge of the door. Common bored locks include deadbolt locks and key-in-knob locks.

bow   That portion of a key that is held between the fingers when the key is being used.

break-away padlock   (frangible padlock) A padlock with a shackle that's designed to be easily broken off. Break-away padlocks are commonly used on fire-safety equipment and on other items that must be readily accessible.

broken key extractor   (broken key puller or key extractor) A tool used to remove a broken piece of key or other foreign matter from the keyway of a lock.

burglar resistant   A term used to describe locks or doors capable of resisting attack by prowlers and thieves for a limited time.

cam   A piece attached to the back of a plug that rotates in conjunction with the plug. In many types of locks, the cam operates the lock bolt. In some cases, such as with many drawer locks, the cam is the lock bolt. Unlike a tang or tailpiece, a cam has no play and operates around a central axis.

cam lock   A lock that has an attached cam that serves as the lock's bolt. Cam locks are often used on cabinets, file cabinets and drawers.

case   (of a cylinder) (often called the housing; sometimes called the shell) The part of a cylinder that houses the plug.

case   (of a lock) The box containing the lock-operating mechanism.

case ward   A ward or obstruction integral to the case of a warded lock.

CDC   abb. Certified Door Consultant (as certified by DHI).

change key   A key that operates only one lock or a group of keyed-alike locks in a series. In a master keyed system, change keys are the lowest level keys.

clip removal tool   (or horseshoe clip removing tool) A tool used to remove retaining clips from an automobile door handle and panel.

CML   abb. Certified Master Locksmith (as certified by ALOA).

combination lock   A lock that may or may not be operated with a key, but can be operated by inserting a combination of numbers, letters, or other symbols by rotating a dial (or dials), or by pushing buttons.

connecting bar   A flat bar attached to the rear of the plug in a rim lock to operate the locking bar mechanism.

control key   A key used to remove the core from a removable core or interchangeable core cylinder.

convenience key   A key that has the same cuts on the top and bottom edges of its blade, and can operate the lock with either side of the blade facing up. A convenience key has cuts on two sides for convenience only, because the lock it operates has only one set of tumblers.

core   The term is sometimes used as a synonym for *plug*; but *core* is also used to refer to the figure 8 shaped unit that can be removed and replaced in interchangeable core cylinders.

corrugated key   A key with pressed longitudinal corrugations in its shank to correspond to a compatibly shaped keyway.

CPL   abb. Certified Professional Locksmith (as certified by ALOA).

CPP   abb. Certified Protection Professional (as certified by ASIS).

current   The flow of electricity. Current is measured in amperes.

cuts   The indentations, notches, and cutouts made in a key blank in order to fit the key to a lock.

cylinder   A unit housed in a case (or shell) and usually containing a plug and tumblers.

cylinder guard   A device designed to protect a cylinder from wrenching; it is usually installed around the rim of a cylinder. Free-spinning cylinder guards spin freely (independent of the cylinder) when being turned.

cylinder lock   A lock having mechanisms operated by a cylinder.

cylinder ring   A collar or washer used under the head of a cylinder.

cylinder screw   The setscrew in the front of a cylinder lock which prevents the cylinder from being turned after installation.

dc   Direct current. For practical purposes, direct current is the type of electricity obtained from batteries.

deadbolt   (or dead bolt) A lock bolt, usually rectangular, that has no spring action, and that becomes locked against end pressure when fully projected.

deadbolt lock   (or deadbolt, tubular deadbolt, tubular dead lock, cylindrical dead bolt) A lock that has a deadbolt and is installed by cross boring two holes in a door.

deadlatch   A lock with a beveled latchbolt that can be automatically or manually locked against end pressure when projected.

deadlock   A lock that projects a deadbolt.

decapper (or de-capper) Pliers that are designed to remove automobile lock caps.

depth key A special key that enables a locksmith to cut blanks made for a special lock according to a code.

derivative code A special numerical code that relates a lock's tumbler arrangement to the depth of the key cuts necessary to operate the lock.

DHI abb. Door and Hardware Institute.

dimple key A key with cuts drilled or milled into the surface of its blade.

disc tumbler (or wafer tumbler, disc-tumbler, disk tumbler) A flat circular or oval-shaped plate with a rectangular hole and one or more side projections. A number of these are used side by side in a disc tumbler lock.

door check A device consisting of a heavy spring and arm coupled to an air or oil cylinder that controls the speed in which a door closes. Also called a door closer.

door reinforcer (or door edge shield, door plate) A U-shaped piece of metal designed to wrap around the edge of a door to strengthen the door.

double-bitted key A key with cuts on both sides of its blade to activate tumblers in a lock.

double-bitted lock A lock that has spring-loaded tumblers from opposite directions.

double cylinder lock A lock that has two cylinders.

drill pin A round pin projecting from the inside of a lock case to receive a barrel key. Also called a barrel post.

drive-in latch A cylindrical latch with a cylindrical knurled ring around its face designed to grip the edge bore of a door. Drive-in latches are commonly used on metal doors.

drivers (or top pins or upper pins) The upper set of spring-activated pins in a pin tumbler cylinder lock. Drivers are usually cylindrical and flat on both ends.

dummy cylinder A nonlocking device that looks like a cylinder, and is used to cover up a cylinder hole.

emergency key A key capable of opening a privacy function lockset even though the door may be locked from inside a room.

escutcheon A plate, either protective or ornamental, containing openings for the controlling members of a lock such as knob, handle, cylinder, keyhole, etc.

fence (also called *the stump*) A piece rigidly attached to the bolt in a lever tumbler lock and, when properly aligned, enters the lever gates to open the lock.

finished key A key that has been cut to fit a lock.

gate The opening in lever tumblers that allows them to pass the post, or fence. Also called gating.

heel (of a padlock shackle) The end of a padlock shackle that is not removable from the case.

IAHSSP abb. International Association of Home Safety and Security Professionals. A trade association for locksmiths, safe & vault technicians, alarm installers, security consultants and others who provide safety- or security-related products or services for or to the residential market.

jamb The inside vertical face of a doorway or window frame.

key blank   An uncut key. Once cut, a key blank becomes a finished key.

key caliper   Any small caliper capable of measuring the height of a cut on a key.

key-in-knob lock   (or knob lock, cylindrical lock) A lock that uses one or two knobs, and has a lock cylinder in one or both knobs.

key plate   A small plate or escutcheon having a keyhole only.

keyway   The opening in a lock mechanism into which the key is inserted.

latch   A door fastening device, usually with a sliding or springbolt.

latchbolt   A beveled springbolt, usually operated by a knob, lever handle, or thumb piece.

layout board   A board with a number of parallel grooves used to hold pin tumbler parts in order when the locksmith is working on the lock.

LED   abb. light-emitting diode. A device that emits light.

lever handle   A horizontal handle for operating the latchbolt of a lock.

lever lock   (or key-in-lever lock, lever handle lock) A lock that has a lever handle.

lever tumbler lock   (or lever lock, lever-tumbler lock) A lock that has a lever tumbler cylinder.

lock   A device that incorporates a bolt, cam, shackle or switch to secure an object—such as a door, drawer, or machine—to a closed, locked, off or on position, and that provides a restricted means of releasing the object from that position.

lockface   A plate that shows in the edge of a door after lock installation.

locking dog   (of a padlock) The part that engages the shackle and holds it in the locked position.

lock pick   (or lockpick, picking tool) Any tool, other than the lock's key, designed to enter a keyway and manipulate a lock's tumblers into a position that allows the plug to be turned.

master key   A key capable of operating several different locks, each lock being operated by its own change key.

mortise   A cavity cut into a door to receive a lock or other hardware. Also the act of making such an opening.

mushroom pins   Mushroom shaped drivers.

nightlatch   An auxiliary lock with a spring latchbolt and functioning independently of, and providing additional security to, the regular lock on the door.

NSLA   abb. National Locksmith Suppliers Association.

ohm   A unit of resistance to electrical current.

panic bolt   A type of lock fitted with a long bar placed horizontally across the inside of a door. Pressure on the bar releases the lock. Used in theaters, schools, and other public buildings, it is also called a panic bar.

P.A.S.S. Key   abb. Personalized Automotive Security System, an anti-theft system made by General Motors. P.A.S.S. Key is the replacement name for the company's Vehicle Anti-theft System (or VATS).

pass key   A master key or skeleton key.

pawl   A tailpiece or cam for an automobile door or deck lock.

pick gun   A tool, often shaped somewhat like a gun, with a blade that's designed to enter a keyway and slap the bottom of tumblers so that they jump into a position that momentarily frees the plug to be rotated. A pick gun can be mechanical or electrical.

pin tumbler lock   (or pin-tumbler lock) A lock that has a pin tumbler cylinder.

pin tumblers   Small sliding pins in a lock cylinder that work against coil springs and prevent the cylinder plug from rotating until the correct key is inserted.

pin tweezers   A pair of tweezers specially designed to handle pin tumblers.

plug   (or core) (of a lock cylinder) The cylindrical mechanism in a lock cylinder that houses the keyway.

plug follower   (or following tool, follow tool) A cylindrical tool used to prevent the drivers and springs from coming out of the upper pin chambers while the plug is being removed from the cylinder case.

plug holder   A fixture used to hold a plug (or plugs) while rekeying or repairing a lock.

plug retainer   A device that secures the plug in a lock cylinder.

post   (of a key) With respect to bit keys, the post is the part of the key that the bit is attached to.

power   The product of current and voltage.

rekey   (or re-key) To modify a lock, usually by changing its tumblers, so that the lock can be operated by a key different from the one that operated it before it was rekeyed.

resistance   Opposition to electrical current flow.

resistor   A component that resists electrical current flow in a dc circuit.

reversible lock   A lock in which the latchbolt can be turned over, so as to adapt it to doors of either hand, opening in or out. This does not apply to beveled front locks and certain cylinder mortise locks.

RL   abb. Registered Locksmith (as certified by ALOA).

SAVTA   abb. Safe and Vault Technicians Association.

shackle   The curved portion of a padlock that passes through the hasp.

shackle spring   The spring inside a padlock that projects the shackle from the padlock case when unlocked.

shank   (of a key) With respect to bit keys, a shank is the part between the shoulder and the bow.

shear line   The space between the housing and the plug of a lock cylinder.

skeleton key   A warded lock key cut especially thin to bypass the wards in several warded locks, so the locks can be opened.

strike box   (or strike reinforcement) A strike plate that incorporates a metal or plastic box that encloses the lock's bolt or latch when in a locked position.

strike plate   (or keeper) A metal plate with one or more cutouts that is installed on a door jamb over one or more corresponding cutouts on the jamb. The cutouts on the strike and jamb are positioned to receive a lock's bolt, latch, or fastener.

tension wrench   (or turning tool, torque wrench, torsion wrench, turning wrench) A tool, usually made of spring steel, used in conjunction with a lock pick. While the pick is moving a lock's tumblers into a position that frees the plug to be rotated, the tension wrench places rotational force on the plug. Although "tension wrench" is older and more commonly used than any of its synonyms, a large minority of locksmiths think the term is technically inaccurate and shouldn't be used by locksmiths.

top pin   (of a padlock) A pin that retains the shackle in a padlock case when unlocked.

torque wrench   See *tension wrench*.

torsion wrench   See *tension wrench*.

tubular key   A key with a hollow cylindrical-shaped blade that has indentations (usually seven or eight) around the rim of the blade. The key is used to operate tubular key locks.

tubular key lock   (or tubular lock, tubular-key lock) A type of pin tumbler lock that has its tumblers arranged in a circle. The lock is often used on vending machines and coin-operated washing machines, and is operated by a tubular key.

tubular lock   See *tubular key lock*.

tubular lock   (or key-in-knob lock, tubular key-in-knob lock) A key-in-knob lock that has two screw posts, one on each side of the lock's central spindle.

tumbler   A small object, usually made of metal, that moves within a lock in ways that obstruct a lock's operation until the proper key or combination moves it into alignment.

turning tool   See *tension wrench*.

UL   abb. Underwriters Laboratories.

VATS   abb. Vehicle Anti-theft System. An electromechanical system that uses a special key to start an ignition. The system is used in many General Motors vehicles manufactured after 1989. VATS is sometimes called *pass-key*.

Volt   A unit of voltage.

Voltage   (also called electromotive force) The force that pushes current. Voltage is measured in volts.

ward   A fixed projection inside a lock that's designed to obstruct unauthorized keys from operating the lock.

warded lock   A lock that relies primarily on wards to prevent unauthorized keys from operating the lock.

Watt   A unit of power.

Index

Illustrations are in **boldface**.